This is an important and much-needed monograph, which offers the first exhaustive attempt to deal with the development of forensic science in late nineteenth and early twentieth-century Britain. It examines forensic science in an interdisciplinary, and broad-ranging, manner to establish the way in which it developed to reflect the needs of the courtroom and the police. It also vitally establishes how its development was presented to the public in the literature of the day.

Keith Laybourn, Diamond Jubilee Professor and Professor of History, University of Huddersfield, UK

The consumer of this book is in for rich fare. Ranging from the idea of the "fact" in law and science, to the rhetorical influence of Sherlock Holmes, to disputes over an in-house police forensic laboratory for London's police in the 1930s, this book ably compounds interpretations of law, science and culture to create a new understanding of scientific criminology. It is an excellent case study in scholarly science studies. You will also watch television crime drama quite differently after reading this book.

Robert Bud, Research Keeper, The Science Museum

This carefully researched book will appeal to historians of crime, policing and criminology, not just those of us interested in the history of forensic medicine and science. Its main strength lies in its skilful weaving of the multiple overlapping trajectories – scientific, medical, political, legal and even literary – from which forensic science emerged as a recognisably "scientific" profession.

Katherine Watson, Senior Lecturer in the History of Medicine since 1500, Oxford Brookes University

A History of Forensic Science

How and when did forensic science originate in the UK? This question demands our attention because our understanding of present-day forensic science is vastly enriched through gaining an appreciation of what went before. *A History of Forensic Science* is the first book to consider the wide spectrum of influences which went into creating the discipline in Britain in the first part of the twentieth century.

This book offers a history of the development of forensic sciences, centred on the UK, but with consideration of continental and colonial influences, from around 1880 to approximately 1940. This period was central to the formation of a separate discipline of forensic science with a distinct professional identity, and this book charts the strategies of the new forensic scientists to gain an authoritative voice in the courtroom and to forge a professional identity in the space between forensic medicine, scientific policing and independent expert witnessing. In so doing, it improves our understanding of how forensic science developed as it did.

This book is essential reading for academics and students engaged in the study of criminology, the history of forensic science, science and technology studies and the history of policing.

Alison Adam is Professor of Science, Technology and Society at Sheffield Hallam University. She is the author of *Artificial Knowing: Gender and the Thinking Machine* and *Gender, Ethics and Information Technology*.

Routledge SOLON Explorations in Crime and Criminal Justice Histories

Edited by Kim Stevenson, University of Plymouth; Judith Rowbotham, University of Plymouth; David Nash, Oxford Brookes University and David J. Cox, University of Wolverhampton

This series is a collaboration between Routledge and the SOLON consortium (promoting studies in law, crime and history) to present cutting-edge interdisciplinary research in crime and criminal justice history through monographs and thematic collected editions which reflect on key issues and dilemmas in criminology and socio-legal studies by locating them within a historical dimension. The emphasis here is on inspiring use of historical and historiographical methodological approaches to the contextualizing and understanding current priorities and problems. This series aims to highlight the best, most innovative interdisciplinary work from both new and established scholars in the field through focusing on the enduring historical resonances to current core criminological and socio-legal issues.

1 **Shame, Blame and Culpability**
Crime and violence in the modern state
Edited by Judith Rowbotham, Marianna Muravyeva and David Nash

2 **Policing Twentieth-Century Ireland**
A history of An Garda Síochána
Vicky Conway

3 **Capital Punishment in Twentieth-Century Britain**
Audience, justice, memory
Lizzie Seal

4 **The Origins of Modern Financial Crime**
Historical foundations and current problems in Britain
Sarah Wilson

5 **Transnational Penal Cultures**
New perspectives on discipline, punishment and desistance
Edited by Vivien Miller and James Campbell

6 **The Police and the Expansion of Public Order Law in Britain, 1829–2014**
Iain Channing

7 **Public Indecency in England 1857–1960**
'A serious and growing evil'
David J. Cox, Kim Stevenson, Candida Harris and Judith Rowbotham

8 **A History of Forensic Science**
British beginnings in the twentieth century
Alison Adam

A History of Forensic Science

British beginnings in the twentieth century

Alison Adam

Routledge
Taylor & Francis Group
LONDON AND NEW YORK

First published 2016 by Routledge

2 Park Square, Milton Park, Abingdon, Oxon OX14 4RN
711 Third Avenue, New York, NY 10017, USA

Routledge is an imprint of the Taylor & Francis Group, an informa business

First issued in paperback 2017

British Library Cataloguing-in-Publication Data
A catalogue record for this book is available from the British Library

Library of Congress Cataloging-in-Publication Data
Adam, Alison.
A history of forensic science : British beginnings in the twentieth century / Alison Adam.
pages cm. — (Routledge SOLON explorations in crime and criminal justice histories ; 7)
Includes bibliographical references.
1. Forensic sciences—Great Britain—History—20th century.
2. Forensic scientists—Great Britain—History—20th century.
3. Criminal investigation—Great Britain—History—20th century.
I. Title.
HV8073.A5276 2016
363.2509—dc23
2015021503

ISBN: 978-0-415-85642-3 (hbk)
ISBN: 978-1-138-30479-6 (pbk)

Typeset in Times New Roman
by Apex CoVantage, LLC

Dedicated to the memory of Evelyn Campbell (1916–2013)

Contents

Figures

Acknowledgements

I should like to thank the University of Salford for granting me a short sabbatical in 2012 to work on this book. Particular thanks are due to colleagues at the Cultural, Communication and Computing Research Institute at Sheffield Hallam University for providing such a supportive research environment where interdisciplinary projects such as this one may flourish. I should like to thank St John's College, Oxford, for granting me a visiting scholarship in the summer of 2011 which provided useful access to much library material. Grateful thanks are also due to the staff of the Griffith Institute, Oxford, for guiding me through Alfred Lucas's Tutankhamun archive materials. I have presented aspects of this research at various seminars and conferences and am grateful for feedback from these, particularly from audiences at the conferences of the British Society for the History of Science. Thanks are also due to the SOLON editors, particularly Judith Rowbotham for helpful suggestions, and to Routledge staff, especially Heidi Lee. I would also like to thank British Pathé for their kind permission to reproduce images belonging to them in Chapter 4 of this book. The love and support of one's family are the bedrock, and I would like to thank my family, Nicol, Sibyl and Craig, for supporting me through yet another big writing project, which I cannot promise will be the last. Craig, in particular, provided a wonderful combination of forensic science expertise, encyclopaedic knowledge of detective fiction and quite a few commas.

Series editor introduction

The volumes in this series contribute to the unashamedly interdisciplinary exercise in which SOLON has engaged since its inception in 2000: something now enhanced by the collaboration with Routledge to present cutting-edge interdisciplinary research in crime and criminal justice history. The focus is on issues which, although rooted in the past, also have a crucial current resonance, and so the volumes reflect on key issues and dilemmas which persist in terms of contemporary priorities.

*****

We are delighted to add this volume to the series, particularly because it adds an unequivocally scientific dimension to the interdisciplinary flavour of those already published while maintaining the focus on criminology and law as examined within a historical perspective. The book's concentration on developments in forensic science is a very welcome theme and one that will further underpin debate and discourse around the core issues critical to criminology and criminal justice studies, and we hope this text will encourage other authors to come forward. There is a real need for this, given the point that Alison Adam herself makes – that this is 'A' and not 'THE' story of forensic science in the UK. There are other narratives which remain to be told, for instance, on the Scottish experience, as the clear and thoughtful summary of the methodological approaches chosen underline.

In avoiding a chronological narrative approach in favour of a thematic one, the text is not bound by the precise chronological constraints which can often distort appreciation and understanding of the respective developmental process. In exploring topics over a relative timeframe, whether long or short, what this volume triumphantly achieves is to challenge popular misconceptions about the nature of forensic science and its purpose. This includes the reality that the content stretches beyond criminality and the criminal justice process. In line with current approaches to the history of science, it shows how complex an area of study forensic science actually is, and how the aspects of forensics which *do* relate to the criminal justice process need to be contextualized and understood in the wider framework of forensic science as a whole. But equally, for the criminologist and the socio-legal scholar, as well as historians of crime, what this book does is provide a cogent and coherent framework for aspects of forensic science that are familiar through existing works which concentrate only on certain elements of the developments in forensic science, such as fingerprinting and poisoning. It also promotes a better understanding of where to locate, and so how to appreciate, the importance of figures as diverse as Conan Doyle and Bernard Spilsbury. Exploring these themes through a focus on the development of laboratories and the less prominent but still significant individuals in forensic science provides a fascinating new lens on the subject. Adam also deals seriously, and innovatively, through the use of detective fiction, with the issue of public receptions and understandings of forensic science as part of a two-way process where popular reactions to forensic science are shown to be intriguingly linked to the perceptions of their role by the scientists themselves. Overall, this is a highly creditable addition to the series which we believe will be useful to a wide range of interested readers.

Introduction

Some years ago, I developed and taught a lecture course on the sociology of forensic sciences and technologies for final-year sociology and criminology undergraduates. It was not difficult to persuade students to enrol. Steenberg argues for the significance of the 'forensic turn' in contemporary culture; my students had certainly turned forensic; like so many of us, they found the 'forensic' label irresistible.[1] I had no trouble gathering resources on contemporary and recent issues: DNA wars, fingerprinting, controversies on the use of statistics in court and the like. The contemporary science–law relationship is well represented in the literature as I describe later in the introduction. However, it was a much harder job to find good material for my introductory lecture, where I charted the rise of forensic science in the UK and where I soon began to realize that the first half of the twentieth century was a crucial, but neglected, period. Although there was excellent material on the history of criminal identification and fingerprinting, and similarly useful material on poisoning and related medical matters, at the time there was a paucity of writing on the history of forensic science, as forensic science, in the UK.[2] So, for that lecture, my teaching resources were limited to two book chapters and two PhD theses.[3] Although still valuable and highly relevant works, they were over twenty years old when I started using them in teaching, and little had been published in the intervening period.

The dearth of suitable historical material provided me with the impulse to conduct my own research, an exciting prospect given that I had a long-standing interest in the history of science which academic disciplinary boundaries had hitherto prevented me from fulfilling. Each year, my historical introduction grew. The lengthening bullet-point list of 'key factors in the development of UK forensic science' was squeezed into an ever-decreasing font size, to the level where it could no longer fit onto one lecture slide. Although I talk of British beginnings and the UK, the story centres on England (I also make reference to Continental and colonial influences). This is deliberate. As a number of scholars have pointed out, the situation in Scotland was somewhat different, and not just because of the different legal system operating there. It was also because of the dominance of the professors of forensic medicine at the universities of Glasgow and Edinburgh and the operation of a scientific laboratory under the control of the police.[4]

Because of the way forensic disciplines grew and specialized in the twentieth century, it is important to acknowledge the influence of medico-legal experts in the development of forensic science in the UK. This includes the most famous of them all, Bernard Spilsbury, the superstar forensic pathologist.[5] Forensic medicine is a much older discipline than forensic science; medical practitioners were present in English courts in the seventeenth and eighteenth centuries.[6] Forensic science did not exist as a separate scientific discipline (or set of disciplines) explicitly called 'forensic science' until well into the twentieth century. Medico-legal specialists undertook a wide range of scientific and technical forensic analyses and convincingly presented these in court cases in the first decades of the twentieth century. For instance, a notorious shooting case from the 1920s showed the involvement of medico-legal experts in forensic ballistics experiments.[7] The fact that medical doctors were undertaking such activities, apparently well outside their professional ambit, now seems rather surprising. However, the range of their forensic activities does not seem so unusual when we realize that a contemporary analytical chemist, an expert in forensic chemistry, also laid claim to forensic ballistics as part of his remit as an analytical chemist.[8] This example points to the importance of charting the specialization of forensic disciplines in the twentieth century. The differentiation of forensic medicine and forensic science is a key part of the development of forensic science and, as the ballistics example shows, other than for clearly medical examples, it was not merely a question of a particular forensic domain naturally belonging to medicine or science. Boundaries and professional delineations had to be negotiated and agreed.

In assembling this material into a coherent narrative, I wish to make it clear that this is 'a history of forensic science in the UK', not '*the* history of science in the UK'. This is not just because of the lack of a substantive historical research tradition, but also because the story is genuinely complex and multi-faceted. I am telling one of the possible stories which could be told, and I am conscious that the relatively few works in this field approach the subject from a somewhat different direction to the path that I take. Ambage's PhD charts the story of the development of the Forensic Science Service in England and Wales from material held in the National Archives.[9] Ward considers a wider range than Ambage, in time as well as subject matter, covering the development of forensic science and forensic medicine. She reveals the importance of Home Office–sponsored scientific posts and emphasizes the role of public analysts who were the chemists in local government service responsible for analyses in pursuit of enforcement of food adulteration laws. This role is often overlooked in forensic histories, given that forensic investigation of murder draws so much popular attention and interest.[10] I agree with Ward that the public analyst as forensic scientist has (still) not received sufficient emphasis. Much of the activity of the forensic scientist in the nascent profession centres on the quotidian. Important as it is, murder is not necessarily the main business of scientists involved in forensic work, however much the influence of TV forensic crime dramas, the so-called 'CSI effect', crime fiction and even the biographies of forensic scientists and those involved in medico-legal work, might lead us to believe.[11] These two PhD theses are, to date, the most comprehensive accounts of the development of the institutional aspects of forensic science in England, and I have drawn upon their insights. Watson's more recent book on the history of forensic medicine provides a comprehensive account of the development of forensic medicine in the UK, Europe and the USA, thereby providing a valuable backcloth for more specific accounts of British forensic science.[12]

In this book, my aim is to understand the historical roots of forensic science over a period which approximately spans the end of the nineteenth century and the first four decades or so of the twentieth century – in other words the time period that sees the making of a profession to which the label 'forensic science' could reasonably be attached. The term forensic science was barely used in the UK until the mid-1930s when the Metropolitan Police Laboratory was opened, with the Home Office network of regional forensic science laboratories soon following, and the publication of a number of texts on scientific aids.[13] Indeed, it could be argued that the UK Home Office invented the term 'forensic science'.

There was a choice of two approaches to this material. One potential approach involves converging on a narrower timeframe, setting to one side the development of expert witnessing, criminology, technological influences, wider forensic careers and fictional forensic science to concentrate on descriptions of the scientific techniques which were developed in the new forensic laboratories. The alternative, which is the path I have taken, is to consider a broad timeframe and the wide spectrum of influences on the development of forensic science in Britain, rather than focusing on detailed descriptions of the science in forensic applications in the 1930s. Of necessity, a discussion of particular scientific techniques has been balanced against an understanding of a very wide range of influences, scientific, criminological, organizational and literary.

My perspective is interdisciplinary, drawing on science and technology studies (STS) and criminology, particularly the history of criminology and especially the history of scientific witnessing and policing. Within STS, the history of science and technology provides the main methodological inspiration. Over the last forty or so years, the idea that science and technology are cultural products, to be studied in the same way as other cultural products, has become the norm in the history of science and technology to the extent that it is no longer necessary to argue for it. This is now an uncontentious view, but it was regarded as quite radical when I was first exposed to it as an undergraduate in the 'Edinburgh School', an encounter from which, fortunately, I never fully recovered.[14] Although trends have evolved in the history of science over the last thirty years, continuing insights centre on the avoidance of technological determinism, or the view that technology drives society or has necessary 'impacts' on society, to offer instead a more balanced view which sees the relationship between technology, science and society as multifaceted. Such an approach exhorts us to treat scientific and technological developments symmetrically, looking for the same kinds of causal explanation of truth and error, of technologies that succeeded and technologies that failed.

For instance, this means that we should be wary of uncritically accepting arguments that forensic techniques based on accepted science are necessarily better than those that were not based on science. Instead, we should consider what was accepted as a reliable, effective technique at the time, without judging it in terms of present-day views on scientific credibility. Despite the affordances of particular technologies and the way that a certain kind of user may be designed into or 'inscribed' in the design of a technology, we can never be sure how a technology or scientific technique will be taken up and used, as human beings are endlessly inventive. Just as with light bulbs, bicycles and Bakelite, so, too, with forensic sciences and technologies.[15] This acknowledges that science and technologies which can be put to use for combatting crime can also be used for criminal ends, and this was a perennial concern. An important tenet of contemporary history of science is that science cannot simply be equated with progress, avoiding what can be seen as a 'Whig' view of history.[16] My discussion of biography and analysis of the relationship of detective fiction and forensic science draws inspiration from the historical 'science and literature' tradition, a flourishing area inhabited mainly by historians of science and English literature academics, related to the broad umbrella of science and technology studies.[17] Hence, I have tried to remain methodologically broad with no intention to push back historiographical boundaries. Rather, in the spirit of 'bricolage', I use what is at hand to build what I hope is a convincing and interesting picture of the many things that contributed to the development of forensic science in Britain in the first part of the twentieth century.[18]

This brings me to a confession on historical purity, both in terms of dates and in terms of sources. I do not stick religiously to a precise time span. Some of what I have to say reaches back to the middle of the nineteenth century and even, in Chapter 1, into the eighteenth century and earlier. I hope the reader will agree that this is reasonable given that many of the concepts and problems with which the scientific expert witnesses of the 1930s were wrestling, such as facts, evidence, probability and how to mount a credible performance in a court of law, were of much older origin. At the other end of the time scale, I have drawn upon sources, particularly biographical material, some of it published in the 1960s, 1970s and 1980s where scientists reflected on their earlier careers in the 1920s and 1930s.[19]

This provokes contemplation of historical sources. I adopt the conventional style of splitting the bibliography into primary and secondary sources, acknowledging that, in some cases, the distinction between primary and secondary sources is artificial, particularly when one is studying the near past. Researching recent history does not offer one the luxury of a neat split between contemporary and much older material. I have placed critical texts, including recent critical biographies, in the 'secondary sources' part of the bibliography and forensic scientists' articles and books, biographies and obituaries in the 'primary sources' section. The distinction is between those texts which are the subjects of research and those texts which are enrolled into the critical analysis. Even so, some texts have resisted clear classification, so I make a plea to the reader that if a reference is not where it was expected in one part of the bibliography, please look for it in the other. I have drawn upon a very wide range of resources, not only because of the proximity of the time period, but also because many of the forensic scientists in this study left no manuscript archives of letters and papers for historians to plunder. To be sure, there are manuscript archives for those two professors of medicine, Smith and Glaister, at the universities of Edinburgh and Glasgow in their respective institutions, and nearby, and there is extensive material on the Home Office initiative in setting up forensic science laboratories in the 1930s in the National Archives.[20] In fact, Smith and Glaister appear in this work more often than I had originally anticipated, such was their influence throughout the UK. However, there is plenty of published material by them and about them, which fosters an appropriate characterization of their roles in the development of forensic science without the need to delve into their archives. The sheer quantity of material from the National Archives on the Home Office initiatives in the 1930s would have prompted a very detailed history of activities that would have affected the space available for considering other important factors.

So the brush is kept fairly broad and paints with a wide range of resources: articles and books written by forensic scientists, including successive editions and translations of textbooks (as the provenance of these and amendments between successive editions are important); biographies and autobiographies; articles, including newspaper articles; official reports and pamphlets, as well as academic critique and commentary. Attempting to cover a wide range of relevant subjects means that, of necessity, for some topics considered briefly, a critical synthesis of secondary materials must be married with primary sources. For instance, the history of fingerprinting has been subject to much popular interest and is described in two excellent scholarly volumes, Cole's *Suspect Identities* and Sengoopta's *Imprint of the Raj*. Hence, I have included a summary of this topic rather than covering well-trodden ground in detail.

The Oxford English Dictionary defines forensic science thus: 'the application of scientific methods and techniques to matters under investigation by a court of law'.[21] In her editorial for a 1989 edition of the *Journal of the Forensic Science Society*, Anita Wonder, an expert in blood spatter analysis, described the relationship between science and the law as 'a marriage of opposites . . . Forensic science is the progeny from a marriage of divergent philosophies'.[22] However, although the relationship between science and law has often been treated as problematic throughout the long history of the marriage and continues to be regarded thus on both sides of the Atlantic, a consideration of the history of their relationship reveals that science and law are far from opposites.[23] Indeed, a number of authors have argued that the marriage of science and law is a fundamental part of the formation of the two disciplines, and many concepts in daily use in both disciplines, such as fact, probability and proof, were constructed through their shared history.[24] We take for granted our understanding of these concepts; the history of how they were made, and the way that they were negotiated and developed, has largely sunk out of trace. The relationship between science and law has a much longer reach than we might at first suppose, the roots of which go back to the seventeenth and sixteenth centuries and probably earlier.[25] This is especially relevant for medical matters in the courtroom, given that medical expert witnesses were probably the earliest form of scientific witness, present from at least the eighteenth century in the UK and much earlier in some European courts.[26]

Hence, although this book centres on developments in the first half of the twentieth century, a brief consideration of the long relationship between science and law, and its implications, is crucial to understanding the connection between the two disciplines and the making of the conditions within which forensic science, or science for (mainly) criminal justice applications, was to develop in the later period. A longer historical perspective also helps us to understand why the law–science relationship has always been regarded as somewhat challenging and shows that later concerns are far from new. In addition, an understanding of the longer historical relationship sheds light upon the initial creation of the role of the scientific expert witness, a role that was to develop significantly with the new forensic scientists of the twentieth century.

We might imagine that contemporary concerns in relation to forensic sciences and technologies, e.g., problems with impartiality and expertise of expert witnesses, 'junk science', keeping accurate records and uniquely identifying individuals, reliably connecting trace evidence to things and people and putting appropriate organizational procedures in place, are recent problems.[27] This is not so. Our understanding may deepen when we know how earlier generations dealt with issues which were important in the development of forensic science from the last decades of the nineteenth century and through the first half of the twentieth century, and which still concern us today.

In sum, the elements which contribute to the development of forensic science in the criminal justice system in the UK include the role of scientific criminology, criminalistics and criminal identification, record management and administrative procedures, scientific policing and detection, the role of scientific aids, the development of a laboratory system and literary influences. If the specific field of twentieth-century British forensic science has been somewhat neglected, then the wider field of the intertwined relationship of law and science, in the UK and elsewhere, has only fairly recently been subject to intellectual scrutiny. When Jones produced her excellent book on expert witnesses some twenty years ago, she argued that a growing literature on the sociology of science was not matched by a literature on the relationship between science and law.[28] Some of the reason for this relative absence, she argued, was due to a propensity to 'black box' both science and the law, as we tend to regard them as bastions of unassailable objectivity.[29] There has also been a certain fragmentation of the area – it is of interest at least to legal and science and technology studies scholars, and possibly scholars from other areas, who sometimes work in quite different intellectual arenas and who may not always know about each other's work, making the topic appear more neglected than it is.[30] However, in the intervening years, these black boxes have been firmly prised open, and there is a burgeoning literature on science and the law, although mainly focused on contemporary rather than historical concerns, much of it located at the intersection of science and technology studies and legal studies.[31]

Chapter 1 considers the way in which legal and scientific terms are imbricated from the eighteenth century and earlier, briefly reviewing the development of science and medicine as professions in the nineteenth century and the way that the nascent professions helped create expert roles. It is not just that law influenced science and vice versa; rather, it is a question of the ways in which the meaning of fundamental terms such as fact, witnessing and probability were mutually created and shaped in law and in science. Shapiro's seminal work on the development of the link between law and science has been particularly useful in providing a backdrop to the later development of this relationship in the twentieth century.[32] These discussions are not purely a historical curiosity. The law–science relationship we have today in places such as England and the USA bears the imprint of centuries-old historical relations. Echoes of the epistemological questions raised by this historical relationship can be detected in the development of forensic science in the twentieth century; these matters still concerned forensic scientists as they reflected on their careers. The professionalization of medicine and science in the nineteenth century was important for enhancing credibility in knowledge production and for using scientific knowledge to enhance expert status. As the examples in second half of Chapter 1 show, forensic scientists and forensic medical experts of the first half of the twentieth century were still wrestling with many of the concepts which developed in science and law from the seventeenth century onward – demonstrating authority, maintaining objectivity and making probabilistic inferences understandable – let alone finding strategies to effectively manage their performance in the courtroom under potentially hostile cross-examination.

Whereas Chapter 1 covers a time scale ranging from the beginnings of expert witnessing in the seventeenth century to the reflections of forensic scientists on their roles in the twentieth century and centres on the courtroom, by contrast Chapter 2 winds the clock back to the nineteenth century (and a little before) and steps out of the courtroom to consider the mechanisms available to identify criminals and criminality and to manage scenes of crime. The influence of scientific criminology is considered, from Europe and from the empire, on the development of forensic science, including measurement and statistics and the importance of evolution. Lombroso, with his measurement of skulls, and Galton, with his composite photographs, are important figures – the latter was much more influential in the UK. Scientific criminology focused on detecting visible measures and markers of criminality on the individual body but had almost nothing to offer to connect the individual criminal to the crime. By contrast, Continental influences, particularly the work of Hans Gross, in articulating the *criminalistic* approach, offered practical suggestions as to how to manage the crime scene, deal with trace evidence and connect an individual to a crime. The first English translations of Hans Gross's influential *Handbuch für Untersuchungsrichter* (*Handbook for Examining Magistrates*) were undertaken by senior British lawyers in Bengal, importing local examples and remaking the book into a kind of handbook of colonial scientific policing which then spread to the UK and USA.[33]

Chapter 3 focuses on technoscience in scientific policing and detection and the contribution of technologies to the development of forensic science. The notorious Crippen case of 1910 is well known; nevertheless it provides a wonderful paradigm example of the enrolment not only of older views on physiognomy but modern technologies of wireless telegraphy, media technologies and forensic toxicology into the business of catching the suspects and trying them in court.[34]

Transport technologies and information and communications technologies, particularly as the latter were applied to Bertillon's anthropometric system and fingerprinting to make them effective identification methods, are an important part of the story. However, it is a sobering thought that Gross's criminalist system, its colonial English-language adaptations, *Bertillonage* and fingerprinting were all significant mechanisms used to foster mistrust and to criminalize nomadic peoples.[35]

Chapter 4 turns the spotlight onto the UK, to include a discussion of the work of UK Home Office's *Scientific Advisory Committee* (1936) and the *Detective Committee* (1938). The latter explicitly advocated scientific aids for criminal detection and was influential in professionalizing the detective role and in fighting for the launch of the forensic science laboratory network of England and Wales.[36] The push for 'scientific aids' in crime detection fostered the division of labour between detective and scientist in managing crime scenes and emphasized the burgeoning view that appropriate collection of trace evidence was a significant policing task to support the work of the new forensic laboratories. The Metropolitan Police Laboratory opened with a fanfare in 1935, but was less than successful in its first decade. The 'scientific aids' movement, as I have dubbed it, was then the British version of criminalistics, arising in response to a perceived need to professionalize and modernize the detective role, to make policing more scientific and to actively bring science to the support of solving crimes.

Chapter 5 considers forensic careers and the projection of self-image in historical context, contrasting the careers of independent scientific expert witnesses with those in the new state forensic science laboratories. The public analyst role was a well-established route into forensic analysis in the UK that developed in the wake of food adulteration Acts.[37] Because this role is often considered under the heading of public health history, this has tended to obscure the contribution of public analysts and analytical chemists to the development of forensic science. More generally, a chemist would regard himself (and it would have been himself) as an analyst if his work involved the chemical analysis of substances to determine their composition for civil or legal reasons, including deliberate or accidental adulteration of substances which could include, but was not confined to, food and drink. This was one way into a consulting career as an independent scientific expert witness.

In this chapter, I ask the question: How do we understand the careers, aspirations and working lives of the early forensic scientists of the 1920s and 1930s when they were so often 'middle-ranking' state scientists who left no personal archive? Some such forensic scientists (and indeed medico-legal specialists) wrote autobiographies. No doubt publishers and authors were well aware that anything to do with murder was profitable. The forensic memoir can be regarded as a branch of the popular, lucrative and well-established detective memoir genre. Forensic memoirs offer clues to the lives of early forensic scientists.[38] With a number of caveats about the limitations of biographies in mind, nevertheless these may be read alongside other popular writings to allow us to gain an understanding of how forensic practitioners presented themselves and their work – in effect how they projected self-image. They were conscious of the need for professional demarcation and were arguing for a distinct professional role for the scientist in forensic work, sometimes in relation to the delineation of forensic science and forensic medicine. Subtle and not-so-subtle pleas for professional status were interleaved with their descriptions of their roles in solving crimes.

The final substantive chapter considers the two-way influence between crime/detective fiction and forensic science and also describes the fascination that the purported relationship has engendered, where the latter is as important as the former. This chapter might seem like something of a change of gear, given that earlier chapters have relatively little to say about literary matters. Nevertheless, I believe that detective fiction was strongly influential in shaping public opinion and in influencing the rhetoric used by those involved in scientific detection to delineate their professional roles. Detective fiction supplies a lens for us to interrogate the history of forensic science. The 'Sherlock Holmes effect' is quite remarkable. Many forensic scientists and medico-legists nod to Sherlock Holmes as an influence – some more than nod, claiming he invented the subject![39] Yet, as more perceptive forensic specialists note in their biographies and other writings, Holmes was not an especially good scientist; his employment of scientific analyses was sketchy at best. Austin Freeman's John Thorndyke was a much better forensic scientist, which some real-life forensic scientists recognized, and this makes him worthy of consideration alongside Sherlock Holmes.[40] Life became woven into fiction in interesting ways. Sir Arthur Conan Doyle donned the mantle of 'Sherlock Holmes' to defend a young man accused of maiming animals. Richard Austin Freeman repeated forensic experiments, even those undertaken originally by a forensic chemist, to enhance the forensic authenticity of his fiction.[41] Authors asked forensic scientists and medico-legal specialists to advise on the scientific aspects of their work; hence, real-life scientists were re-imagined in fictional form.

Modern commentators work hard to find the science in Sherlock Holmes, but with limited success. Rather, it is the 'scientific method' which Holmes applied which is the clue to his appeal as the basis of an origin story, a primordial soup from which modern forensic science is held to evolve – a powerful myth even if things did not quite happen like that. This reinforces the overarching theme of the book, namely the use of the term 'scientific' in relation to the detection of crime, what 'scientific' was taken to mean and how a scientific approach in scientific criminology, criminalistics and organization and a wider technoscience of information, communications, media and transport technologies and scientific aids to criminal investigation, all shaped the development of forensic science in Britain in the first part of the twentieth century.

Notes

1 L. Steenberg, *Forensic Science in Contemporary American Popular Culture: Gender, Crime and Science*, Abingdon: Routledge, 2013.

2 See I. Burney, *Poison, Detection, and the Victorian Imagination*, Manchester: Manchester University Press, 2006; K. D. Watson, *Poisoned Lives: English Poisoners and*

Their Victims, London and New York: Hambledon and London, 2004; K. D. Watson, 'Medical and chemical expertise in English trials for criminal poisoning, 1750–1914', *Medical History*, 2006, 50: 373–390; K. D. Watson, *Dr Crippen*, Kew, Richmond: The National Archives, 2007; S. A. Cole, *Suspect Identities: A History of Fingerprinting and Criminal Identification*, Cambridge, MA: Harvard University Press, 2001; C. Sengoopta, *Imprint of the Raj: How Fingerprinting Was Born in Colonial India*, Basingstoke and Oxford: Pan, 2003; for writing on the history of poisonings and fingerprinting, respectively, published before 2010.

3 N. Ambage, and M. Clark, 'Unbuilt Bloomsbury: medico-legal institutes and forensic science laboratories in England between the wars' in M. Clark and C. Crawford (eds), *Legal Medicine in History*, Cambridge: Cambridge University Press, 1994, 293–313; M. A. Crowther, and B., White, *On Soul and Conscience. The Medical Expert and Crime. 150 Years of Forensic Medicine in Glasgow*, Aberdeen: Aberdeen University Press, 1988, Chapter 6; N. Ambage, *The Origins and Development of the Home Office Forensic Science Service, 1931–1967*, unpublished PhD thesis, Lancaster University, 1987; J. Ward, *Origins and Development of Forensic Medicine and Forensic Science in England, 1823–1946*, unpublished PhD thesis, The Open University, 1993.

4 Crowther, and White, *On Soul and Conscience*; N. Duvall, *Forensic Medicine in Scotland, 1914–39*, unpublished PhD thesis, University of Manchester, 2013.

5 A. Rose, *Lethal Witness: Sir Bernard Spilsbury Honorary Pathologist*, Chalfont, Stroud: Sutton, 2007.

6 K. D. Watson, *Forensic Medicine in Western Society: A History*, Abingdon: Routledge, 2011, p. 9.

7 This is the 1926 Merrett case in Edinburgh and is discussed in Chapter 1. See S. Smith *Mostly Murder*, London: Harrap, 1959, p. 143.

8 A. Lucas, *Forensic Chemistry*, London: Edward Arnold, 1921.

9 Ambage, *The Origins*.

10 Ward, J., *Origins and Development*.

11 S. A. Cole and R. Dioso-Villa, 'Investigating the "*CSI* Effect": Media and Litigation Crisis in Criminal Law', *Stanford Law Review*, 2009, 61(6): 1335–1373. The 'CSI effect' is the postulated effect that the portrayal of forensic sciences and techniques in crime dramas such as *CSI* (*Crime Scene Investigation*) may have in court proceedings and on the public.

12 Watson, *Forensic Medicine*.

13 HMSO, *Home Office: Scientific Aids to Criminal Investigation: Instructional Pamphlet for the Use of Police Officers*, His Majesty's Stationery Office, 1936. For articles see, e.g., F. G. Tryhorn, 'Scientific aids in criminal investigation Part I', *The Police Journal*, 1936, 9 (1): 33–41, F. G. Tryhorn, 'Scientific aids in criminal investigation. Part II Searching at the scene of crime', *The Police Journal*, 1936, 9 (2): 152–160.

14 See S. Sismondo, *An Introduction to Science and Technology Studies*, Malden, MA: Blackwell Publishing, 2010. The 'Edinburgh School' refers to the approaches towards the social, historical and philosophical study of science and technology that flourished in the Science Studies Unit at the University of Edinburgh in the 1970s and 1980s.

15 For classic works in science and technology studies which broadly conform to this approach, see W. E. Bijker, *Of Bicycles, Bakelites, and Bulbs: Toward a Theory of Sociotechnical Change*, Cambridge, MA: MIT Press; W. E. Bijker, T. P. Hughes and T. J. Pinch (eds), *The Social Construction of Technological Systems: New Directions in the Sociology and History of Technology*, Cambridge, MA: MIT Press, 1987; D. Bloor, *Knowledge and Social Imagery*, London: Routledge & Kegan Paul, 1976;

D. MacKenzie and J. Wajcman, J. (eds), *The Social Shaping of Technology: How the Refrigerator Got Its Hum*, Milton Keynes: Open University Press, 1985.

16 H. Butterfield, *The Whig Interpretation of History*, London: G. Bell and Sons 1950 (original edition 1931).

17 For recent surveys of the science and literature area, see C. Sleigh, *Literature and Science*, Houndmills, Basingstoke: Palgrave Macmillan, 2011; M. Willis, *Literature and Science: A Reader's Guide to Essential Criticism*, London and New York: Palgrave, 2015.

18 For 'bricolage' in academic research, see J. L. Kincheloe and K. S. Berry, *Rigour and Complexity in Educational Research: Conceptualizing the Bricolage*, Maidenhead: Open University Press, 2004.

19 E.g., see J. B. Firth, *A Scientist Turns to Crime*, London: William Kimber, 1960; H. J. Walls, *Expert Witness: My Thirty Years in Forensic Science*, London: John Long, 1972.

20 See Duvall, *Forensic Medicine in Scotland* for details of the Edinburgh and Glasgow archives. See Ambage, T*he Origins and Development* for details of material in the National Archives.

21 See definition at: http://www.oxforddictionaries.com/definition/english/forensic-science?q=forensic+science, accessed 1 May 2015. Many similar definitions of forensic science emphasize the use of scientific analyses in fighting crime and its presentation in court.

22 A.K.Y. Wonder, 'Science and law, a marriage of opposites', *Journal of the Forensic Science Society*, 1989, 29 (2): 75–76, p. 75.

23 P. C. Giannelli, 'The 2009 NAS Forensic Science Report: A literature review', *Criminal Law Bulletin*, 2012, 48 (2): 378–393; National Research Council, National Academy of Sciences (NRCNAS), Strengthening Forensic Science in the United States: A Path Forward, Washington DC, National Academies Press, 2009.; Science and Technology Committee; 'Forensic science on trial: Seventh report of session 2004–05', Report, House of Commons Committee, 2005, at http://www.publications.parliament.uk/pa/cm200405/cmselect/cmsctech/96/96i.pdf, accessed 12 January 2015.

24 B. Shapiro, *Probability and Certainty in Seventeenth-Century England: A Study of the Relationships Between Natural Science, Religion, History, Law and Literature*, Princeton, NJ: Princeton University Press, 1983; B. Shapiro, *Beyond Reasonable Doubt and Probable Cause: Historical Perspectives on the Anglo-American Law of Evidence*, Berkeley and Los Angeles: University of California Press, 1991; B. Shapiro, *A Culture of Fact: England 1550–1720*, Ithaca, NY and London: Cornell University Press, 2000; C.A.G. Jones, *Expert Witnesses: Science, Medicine, and the Practice of Law*, Oxford, Clarendon Press, 1994.

25 Shapiro, *Probability and Certainty*; Shapiro, *Beyond Reasonable Doubt*; Shapiro, *A Culture of Fact*; Jones, *Expert Witnesses*.

26 Jones, *Expert Witnesses*, especially Chap. 6 and p. 20; T. R. Forbes, *Surgeons at the Bailey: English Forensic Medicine to 1878*, New Haven, CT: Yale University Press, 1985.

27 P. Huber, *Galileo's Revenge: Junk Science in the Courtroom*, New York: Basic, 1991.

28 Jones, *Expert Witnesses*.

29 Ibid.

30 For instance, Tal Golan argues that the history of scientific expert testimony has long been overlooked by historians of science and historians of law; hence, there is relatively little scholarship on the subject. However, the work of Barbara Shapiro, who is the main apologist for the thoroughly intertwined historical relationship between important legal and scientific concepts, is significant in this area. See T. Golan, 'The

history of scientific expert testimony in the English courtroom', *Science in Context*, 1999, 12 (1): 7–32.

31 See, for instance, J. D. Aronson, *Genetic Witness: Science, Law, and Controversy in the Making of DNA Profiling*, New Brunswick, NJ: Rutgers University Press, 2007; D. S. Caudill, 'Arsenic and old chemistry: Images of mad alchemists, experts attacking experts, and the crisis in forensic science', *Villanova University School of Law Working Paper Series, Paper 136, 2009* Available at http://digitalcommons.law.villanova.edu/cgi/viewcontent.cgi?article=1140&context=wps, accessed 3 January 2015; Cole, *Suspect Identities*; G. Edmond (ed.), *Expertise in Regulation and Law*, Aldershot: Ashgate, 2004; R. Hindmarsh, and B. Prainsack, (eds), *Genetic Suspects: Global Governance of DNA Profiling and Databasing*, Cambridge: Cambridge University Press, 2010; S. Jasanoff, *Science at the Bar: Law, Science and Technology in America*. Cambridge, MA: Harvard University Press, 1995; M. Lynch, S. A. Cole, R. McNally and K. Jordan, *Truth Machine: The Contentious History of DNA Fingerprinting*, Chicago, IL, and London: Chicago University Press, 2008; J. L. Mnookin 'Idealizing science and demonizing experts: An intellectual history of expert evidence', *Villanova Law Review*, 2007, 52 (101): 763–801; R. Williams and P. Johnson, *Genetic Policing: The Use of DNA in Criminal Investigations*, 2008, Cullompton: Willan Publishing.

32 Shapiro, *Probability and Certainty*; Shapiro, *Beyond Reasonable Doubt*; Shapiro, *A Culture of Fact*.

33 H. Gross, *Handbuch für Untersuchungsrichter als System der Kriminalistik*, 2 volumes, Munich: J. Schweitzer Verlag, 1893; J. Adam and J. C. Adam, *Criminal Investigation A Practical Handbook for Magistrates, Police Officers and Lawyers*, Madras: A. Krishnamachari, 1906; J. Adam, and J. C. Adam, *Criminal Investigation A Practical Handbook for Magistrates, Police Officers and Lawyers*, London: The Specialist Press, 1907 (reprint of 1906 edition); J. C. Adam, *Criminal Investigation A Practical Textbook for Magistrates, Police Officers and Lawyers*, London: Sweet and Maxwell, 1924. All English editions included this text on the title page: 'Translated and adapted from the System der Kriminalistik of Dr Hans Gross, Professor of Criminology in the University of Prague. By John Adam, M.A., Barrister-at-Law, Crown and Public Prosecutor, Madras and J. Collyer Adam, Barrister-at-Law, Advocate, High Court, Madras'.

34 Watson, *Dr Crippen*.

35 See Cole, *Suspect Identities* on fingerprinting and Bertillon's system. See D. M. Vyleta, *Crime, Jews and News: Vienna 1895–1914*, New York and Oxford: Berghahn, 2007.

36 HMSO, *Home Office: Report of the Advisory Committee on the Scientific Investigation of Crime*, London: His Majesty's Stationery Office, 1936; HMSO, *Home Office: Report of the Departmental Committee on Detective Work and Procedure*, London: His Majesty's Stationery Office, 1938.

37 G. Taylor, *Forensic Enforcement: The Role of the Public Analyst*, Cambridge: RSC Publishing, 2010.

38 H. Shpayer-Makov, 'Explaining the rise and success of detective memoirs in Britain', in C. Emsley and H. Shpayer-Makov (eds) *Police Detectives in History, 1750–1950*, Aldershot: Ashgate, 2006, 103–133.

39 See S. O. Berg, 'Sherlock Holmes: Father of scientific crime detection', *Journal of Criminal Law and Criminology*, 1970, 61 (3): 446–452, for what is possibly the most enthusiastic endorsement of the claim that Sherlock Holmes invented forensic science.

40 N. Donaldson, *In Search of Dr Thorndyke: The Story of R. Austin Freeman's Great Scientific Investigator and His Creator*, Bowling Green, OH: Bowling Green University Popular Press, 1971.

41 Ibid.

1 The relationship between science and law

Expert witnesses in the courtroom

Introduction

We begin by considering how the relationship between science and law influenced the development of scientific epistemology and how early echoes of this relationship continued to manifest themselves in the 'warring experts' in the courtroom phenomenon of the Victorian period and well into the twentieth century, where the new forensic expert witnesses grappled with their courtroom performances. A number of epistemological distinctions, important for the law and for science, were developed from the seventeenth century onwards. These include the concept of fact and evidence (especially physical evidence), the concepts of certain and probable knowledge (in other words, probability), witnessing and especially scientific expert witnessing. Scientific and legal epistemological categories are embedded within one another; they are mutually constitutive.[1] The notion of an expert witness was, to a considerable extent, developed in the seventeenth century where the man of science came to be seen as a reliable witness.[2] Proof and probability are central to the law and to science, and they are made and shaped in both places.[3] After characterizing these developments, we turn to the development of the expert witness role and the significance of the professionalization of medicine and science in the nineteenth century.

Fundamental epistemological categories, which have become naturalized to the extent that their origins have disappeared from our collective consciousness, were used to claim authority and status, and eventually in the nineteenth century, would become critical markers of professionalization. The ability to assert authority over knowledge production of the natural world was an important element in the professionalization of science in the nineteenth century. Such a strategy was also evident in the professionalization of medicine in the same period, where attempts to gain autonomy and status were enhanced by claims for a scientific basis to medical knowledge. The professionalization of medicine and science were important elements helping to consolidate the expert witness role. The second half of the chapter turns to battles for authority that expert witnesses found themselves in the midst of in court, with the middle-to-late nineteenth century as the nadir of the battling experts period. Such problems did not disappear in the twentieth century. Hence, the strategies that later medico-legal experts and forensic scientists adopted to achieve authority in the courtroom under cross-examination and to deal with centuries-old epistemological categories of facts, evidence and probabilities provided a direct lineage from the seventeenth to the twentieth century.

There were tensions between the certainty that scientific knowledge appeared to promise in the hands of credible scientific witnesses and the demands of the adversarial legal system. This meant that scientists acting as expert witnesses had to be adept at presenting evidence in court, holding their nerve in the face of hostile cross-examination. Claiming and maintaining authority in court was a major issue, not just in terms of the ability of the individual scientist's credibility, but also in terms of the accepted credibility of a scientific technique. Handwriting evidence is a paradigm example of a forensic technique which eventually became accepted in court. For many, scientists included, the prospect of battling experts ranged on opposite sides of the courtroom was regarded as problematic. For, if scientific knowledge was objective and factual, as centuries of epistemological development in law and science had taught us to believe, how could scientists be put in such a situation of profound disagreement? A notorious poisoning case from the nineteenth century provides an example of the battle of experts at its most problematic.

Scientists who were acting as expert witnesses had to mount a performance in court, and it is clear that whereas some relished this and had developed appropriate supporting strategies, for others the performance was a difficult, and not entirely welcome, part of the job. On the other hand, the alternative, namely presenting written evidence without a court appearance, had its pitfalls too. Some medico-legal practitioners and forensic scientists were known to be star performers in the courtroom – indeed, some of the rising generation of forensic scientists recruited to the new roles opening up in the 1930s were head-hunted, at least in part, for their courtroom skills. But others felt that the need to perform under potentially hostile courtroom conditions was something of a contrast to their detailed, expert scientific work and developed strategies to cope with that contrast, strategies which they sometimes shared in popular and biographical works.

Facts, values and opinions

'In this life, we want nothing but Facts, sir; nothing but Facts!'[4] Thus spoke Thomas Gradgrind, Dickens's utilitarian teacher in *Hard Times*. In his attempts to pour facts to the brim of the vessels of his pupils' minds, Gradgrind was unlikely to have reflected that the concept of the 'fact' was a relatively recent construct. And if he did reflect on the origin of 'facts', given the marvels of Victorian science and engineering, he might have supposed that the idea of a fact came from scientific observations of the natural world.

For instance, even if we might disagree about whether some particular event or thing is itself a 'fact', at the beginning of the twenty-first century, at least in everyday terms, we are unlikely to regard the concept of 'fact' as an epistemological category as particularly contentious. Even post-modern academics whose careers are built on a denial of the concept of fact, in shopping, cooking and catching trains, tend to act as though the world is full of facts. So ingrained is our understanding of facts that we have difficulty imagining a time when we did not have them. Fact, arguably the central epistemological concept of modern science, derives originally from the law, rather than science or, at least, derives from the intertwining of science and law.[5]

It was within law rather than natural sciences that the concept of 'fact' and the concomitant concept of 'witness' emerged because, of course, we cannot have facts without acknowledged, competent people to testify that something is a fact. Shapiro argues that ' "fact" does not begin with natural phenomena and was, if anything, a rather late arrival in natural philosophy, having become a well-established concept elsewhere before it was adopted by the community of naturalists'.[6] 'Fact' and 'matters of fact' developed in law and were then taken up in other disciplines to the extent that they became part of general culture and intellectual life in the seventeenth and early eighteenth centuries in Britain. The Royal Society, founded in 1660 and still the most venerable scientific society in the UK, did much to develop the concept of fact into an epistemological category in science, and much of its early work as the newly established body for scientific research in the UK was involved with describing and cataloguing natural events and facts.[7]

Francis Bacon (1561–1626) was an important influence in the work of the early Royal Society. Indeed, the Baconian programme of developing scientific knowledge through exhaustive 'fact' gathering was influential in the early Royal Society's scientific programme. As a lawyer, historian and man of science, Bacon was a leading figure in the transfer of 'fact' into the natural sciences alongside the idea that facts were things in the natural world that could be collected and reported by a reliable witness. But the concept of 'fact' was fluid and was far from immutable. Disciplines tended to reinvent the idea of fact finding to suit their own needs and requirements.[8]

So, the idea of 'fact' evolved in legal settings, was quickly transported into other areas and gradually became part of intellectual life by the end of the seventeenth century. In law, a 'fact' referred to a human activity which could be known about or directly witnessed.[9] 'Matters of fact' were for juries to decide; 'matters of law' were the province of judges. It was the familiarity with determining facts that made it so easy to move fact from law into other domains. In English law, 'fact' had two related meanings – the more general meaning of any human act and the more specific meaning of the actual act which constituted the crime which was being tried in the court of law (we still use the term 'after the fact'). Fact meant act, so the concept of fact tended to have a more active meaning in the legal realm than it did in its early scientific manifestation, where the great Baconian activities of gathering and classifying the materials of a somewhat passive natural world prevailed.

The idea that one could reliably know about people precedes the idea that one could reliably know about things. The development of the legal system in England then was a place where 'juries, judges, witnesses, and counsel participated in a process that was designed to produce "morally certain" verdicts in "matters of fact" '.[10] Although first-hand perceptual evidence was desirable, it was recognized that if one could not have first-hand evidence, then a judgment had to be made about the credibility of a witness. Acknowledging Shapin and Schaffer's research on the gentlemanly nature of scientific values of trust in making a scientific culture based on fact and gentlemanly status for credible witnessing, Shapiro nevertheless sees the development of the idea of fact as a more general feature of seventeenth-century intellectual life, arising at around the same time in several disciplines, including history, theology and, in particular, law. Rather than gentlemanly discourse based on social position promoting 'fact', especially scientific fact, Shapiro argues that it was the concept of fact, already in common use, that promoted their appropriately 'gentlemanly' modes of interaction. Ordinary people, as long as they had no conflict of interest, were perfectly able to decide matters of fact; the ability to witness was, therefore, not just confined to those of gentlemanly status.[11]

> Lawyers and virtuosi shared an emphasis on truth, an insistence on fact over fiction and imagination, a preference for firsthand and credible witnessing, and a rhetoric of impartiality. The courtroom and the rooms of the Royal Society shared a great deal.[12]

It was the attention paid to observation and experimentally produced facts which marked out the new experimental science of the Royal Society where sceptical chemist Robert Boyle was the exemplar of the new man of science. As with legal facts, scientific facts were established by witnesses 'whose testimony would be evaluated on the basis of a set of legally derived criteria of credibility, such as opportunity, ability, probity, skill, fidelity, status, experience, and reputation'.[13] Importantly, Shapiro notes that models of witnessing facts already established in law and history were adopted by the natural scientists, and this helps to explain the legal language in which such concepts were expressed. Hence, scientific virtuosi, as seventeenth-century scientists were known, emphasized witnessing, impartiality and cautiousness, and they did not need to explain their use of the term 'fact', as it was already in common use in other walks of life. Gentlemanly status may have been useful for enhancing credibility, but it was by no means essential, as expertise was increasingly seen as important, so scientific experts of a 'middling rank' were as credible as gentlemen.[14]

Fast forward to the end of the nineteenth century and first decades of the twentieth, and we see plenty of men of science of a 'middling rank' attaining the status of credible scientific witness. They were helped by the achievement of the distinction between facts and values and the control of science over the making of facts which had been accomplished by the middle of the nineteenth century. The distinction between fact and value was a major concern in law and science over the last 200 or more years; this separation, a historically contingent split which we now take for granted, fosters the creation of the expert witness role.[15] One of the major successes of modern science was to achieve this split and to attach itself firmly to the idea of value-free, fact-based science. This meant that men of science, like men of law, were able to present their knowledge-making activities as neutral, above political and social concerns, and as described later in the chapter, the ability of twentieth-century forensic scientists and pathologists to portray themselves as objective and neutral, presenting facts and their expert opinion on such facts, was crucial to their success in the courtroom. 'Men of science and men of law have thus come to enjoy an unrivalled position as diviners of the facts. Their findings are held to be the result not of personal whim but of impersonal judgement.'[16]

One of the special features of expert witnesses is that, by dint of their specialist training and expertise, they are permitted to give their opinions on the basis of presented facts (as opposed to presenting facts, which is the business of ordinary witnesses), thus forming causal explanations of past events.[17] Although referring to expert witnessing later in the twentieth century, Roger Smith's analysis of the ways in which expert witnesses must attend to the division between facts and opinions is, nevertheless, highly relevant to earlier expert witnessing, the only substantive difference being that, by the time of his study, written witness statements in court were much more common. Of course, that experts are able to give an opinion demonstrates that facts do not speak for themselves and need context and explanation.[18] Expert witnesses learn to separate facts from opinions in their evidence so that non-specialists can understand them. Smith argues that most forensic scientists and pathologists do not think there are problems distinguishing fact from opinion, believing that competent observers would agree with the same facts. It is a sign of professionalism to report facts in such a way that colleagues would agree with the stated facts.[19] However, the experts in his study agreed that there would be valid differences in opinion as to their explanation of likely causes. Despite this, they believed that if they had the opportunity to discuss their opinions freely with one another, they would almost always agree. In other words, they had the means to transport opinions into facts given the opportunity, although this process would naturally involve probabilistic statements. So agreement could be reached by a reasonable process, but it was the adversarial system to a significant extent which maintained this division. 'It follows, in their view that it is the legal setting which imposes a sharp separation between fact and opinion and the opening up of shared likelihood statements into unshared statements of opinion.'[20] So the prevailing view amongst the expert witnesses in Smith's study was that disagreement as to facts was not generally valid, but differences in opinion were, and it was largely the adversarial system that drove a wedge between facts and opinions, preventing experts from negotiating and agreeing opinion.

Probability – the moral certainty of knowledge

Closely associated with the construction of facts, the split between facts and values and the split between facts and opinions is the question of probabilistic statements. As the title of Shapiro's pivotal work, *Probability and Certainty*, suggests, the definition of probability is intimately linked to the definition of certainty. Importantly, there is a moral dimension to certainty.[21] Law, science, religion, history and many other aspects of social life held 'a common set of assumptions about the nature of truth, the methods for attaining it, and the degree of probability or certainty that might be attributed to the findings produced by those methods'.[22] By the close of the seventeenth century, knowledge was seen as falling along a spectrum from certainty and fact, or 'moral certainty' at one end, through highly probable, probable to conjecture and opinion.[23] Indeed, the shock of seeing scientists fight over conflicting opinions in court in the nineteenth and twentieth centuries emphasizes continuing expectations of the moral dimension of certain knowledge and the guardians of certain knowledge, that is, men of science.[24] No longer were absolute facts the only form of knowledge; probabilistic reasoning gradually became accepted. For science, the concentration on empirical enquiry brought about this move, and this meant that empirical sciences could be seen as separating from mathematics in their quest for certainty.[25]

The question of making true statements about things one could not observe directly was a central concern for scientific expert witnesses who had to negotiate the validity of testimony and authority. This issue was of key importance in the seventeenth century, not least of all for historians making true statements about the past. The meaning of the term 'probability' altered in the seventeenth century to acquire a meaning relating to scientific knowledge: 'an adequate degree of evidential support for a claim that was not *certainly* true'.[26] Handling probability and making sure that probabilistic reasoning with regard to evidence was rendered intelligible has remained a headache for scientific expert witnesses to the present day. Later in this chapter, we can observe the frustration of mid-twentieth-century forensic scientists in relation to presenting evidence which is probabilistic or not completely certain.

The moral dimension remains important. We would be somewhat puzzled if a contemporary statistician were to talk about probabilistic reasoning in terms of moral certainty nowadays. Yet a number of statements as to the moral qualities of knowledge, truth and experience, and hence the moral qualities of forensic specialists, are to be found in the writings of forensic scientists – there are a number of examples later in this chapter. Smith found a strong moral commitment to the 'truth' in his study – such a commitment was regarded a personal quality developed by being part of the scientific profession.[27] '[T]he expert's personal qualities enable a scientifically trained individual to operate with integrity in an adversarial setting which does not share the same value of commitment to "the truth".'[28]

The historical development of the scientific witness

The historical significance of witnessing in the science–law relationship is acknowledged by science and technology studies and legal scholars, and is still centre stage in contemporary debates on forensic sciences. Witnessing, including importantly expert witnessing, is a central component in the Anglo-American legal system. Shapin and Schaffer and also Shapiro argue for the credibility of the seventeenth-century man of science to witness the truth of a scientific fact.[29] Although this is not how we now frame the relationship between scientists and facts in the contemporary world, nevertheless, there are times when the question of scientific witnessing was brought sharply to the attention of the court, as shown in the Victorian poisoning case below, the concerns of Victorian scientists and even in the court appearances of twentieth-century forensic specialists.

Just as the historical development of facts, evidence and probability were historical achievements, so, too, was the role of the scientific witness; Robert Boyle (1627–1691) is regarded as the founding father. Boyle is widely acknowledged to be a central figure in devising an experimental science where appropriately credentialled others could witness and agree to the facts of a scientific phenomenon, signalling a key feature of Enlightenment thinking, namely the move away from men of science seeing themselves purely as philosophers, towards becoming active participants in empirical measurements of natural phenomena.[30] In relation to Boyle's famous pneumatic experiments with an air pump, Shapin and Schaffer characterize his contribution to the business of establishing matters of fact in scientific experiments and how they were to be witnessed in terms of three technologies, or '*knowledge-producing tools*':

> [A] *material technology* embedded in the construction and operation of the air-pump: a *literary technology* by means of which the phenomena . . . were made known to those who were not direct witnesses; and a *social technology* that incorporated the conventions experimental philosophers should use in . . . considering knowledge-claims.[31]

Scientific facts were established by witnessing experiments which could either be done directly – a *social technology* – or by means of a written document, the scientific paper – a *literary technology*. Either way, a set of conventions had to be developed and agreed to, and Boyle was a central figure around which agreed techniques crystallized. Readers of his experimental essays became 'virtual witnesses'; it was therefore possible to expand considerably the number of witnesses to a matter of fact as it was demonstrated by a scientific text; a witness did not have to be physically present as long as the agreed-upon protocols were followed.[32] His contribution shows that the development of making and agreeing to scientific facts depends as much on agreeing on appropriate witnessing technologies and agreeing on the credibility of the witnessing agent as on the material organization of scientific experiments. This has remained important throughout the long history of the law–science relationship. It is never just a question of presenting the results of a laboratory experiment 'cold' in court. Scientific facts do not speak for themselves in court; there has to be a considerable wrap-around of accepted processes, explanation, education and convincing. This is related to the way in which scientists agree on the conventions for what counts as a repeated scientific experiment, part of what Lynch et al. term 'administrative objectivity', the agreed regulatory regime of protocols necessary to agree on the scientific status of evidence.[33]

Although many of these have filtered away from our collective cultural consciousness, then as now, elaborate witnessing procedures had to be in place for a scientific result to be accepted. One of these, the reference to an accepted scientific text, is of particular interest in the forensic context. By the end of the first decades of the twentieth century, there were several textbooks on forensic medicine and forensic science and two English-language editions of Gross's *Handbuch für Untersuchungsrichter* (Handbook for Examining Magistrates).[34] Notably, Alfred Swaine Taylor's (1865) *The Principles and Practice of Medical Jurisprudence* reached its twelfth edition with its centenary.[35] Successive editions were edited by notable medico-legal specialists; indeed, becoming an editor of an edition of *Taylor's Principles* was, in itself, a highly visible badge of scientific and medical authority. Each time a new edition of *Taylor's Principles* was published, another 'virtual witness' added his name as editor; thus, reflexively, the book gained additional credence and the editor also gained authority by being sufficiently eminent to be an editor of such a venerable work, a 'snowballing' effect. It was all very well, being the author or editor of an authoritative work, but any such author had to ensure that the authority of the work was kept under control. The virtual witness of such a scientific text could have his work turned against him in a trial. One trick which legal counsel could try to play against the expert witness was to quote the textbook which the expert had written or edited as a scientific authority and try to get the witness to disagree with himself. Making the expert apparently contradict what he had said in print thus undermined the authority of his evidence *and* the authority of the work.[36]

Shapin argues that for seventeenth-century society, direct experience was regarded as the most important thing for agreeing to factual knowledge.[37] If you were unable to have direct experience, then you had to rely on the testimony of others. You could rely on their testimony if you trusted them, if they had authority. Witnessing had a moral dimension. The scientific community imported aspects of legal witnessing into its model of the 'ideal scientific observer and reporter'.[38] Skill, experience, expertise and integrity were important criteria. Expertise was not confined to high-ranking individuals; practical skills and membership of an appropriate corporation or guild was seen to be as important as academic qualifications.[39]

Robert Boyle may seem to be a very long way from the analytical chemists, public analysts, forensic scientists and medico-legal experts of the first part of the twentieth century, but although his scientific witnessing achievements have become so well accepted in science that they are below the epistemological radar, they still stand guard over later scientific analyses. The new forensic scientists of the twentieth century may have been working in a different scientific environment, nevertheless they are direct heirs of the system of scientific witnessing set in train by Boyle and his contemporaries, which has been maintained and developed over the centuries.

The separation of witnesses and juries

Initially, juries and witnesses were not separate; the achievement of the separation of the role of jury and witness was an important part of developing the role of the witness and, especially, the expert witness.

Early juries were composed of witnesses; if you lived in a community where everyone knew everyone else, then separate witnesses were not deemed necessary.

Gradually, juries became judges of facts rather than knowers of facts, i.e., *no longer* witnesses, by the end of the sixteenth century. The early modern jury was selected from people who knew the facts of the case or who had some expertise. Juries could seek assistance from jurors with some specialist knowledge; court experts and specialist juries of craftsmen were used to decide craft disputes, so their combined knowledge was used as a basis for a verdict.[40] In the early modern period, a distinction between juror and witness was progressively being made, and when juries stopped being constituted from a group of direct witnesses, the role of the modern expert witness gradually emerged. So the structural changes in the jury system were crucial in making the expert witness role, as much as the development of the new mechanical philosophy, in the hands of Boyle and the Royal Society.

From the sixteenth to the eighteenth century, expert witnesses gradually changed from being advisers to the court towards being partisan expert witnesses and, at the same time, they were permitted to give inferences as part of their testimony.[41] In the earlier part of this period, disagreement amongst experts was not encouraged, and the system of court experts and special juries gave relatively little scope for public dispute.[42] But things had begun to change by the end of the eighteenth century, and in the nineteenth century courtrooms were beginning to witness experts set against each other. Although it was accepted that experts would have to give opinions on causal matters, not just facts, increasingly lawyers regarded expert evidence with a level of suspicion, given its partisan qualities.[43] So, over a period of centuries, the move was made from juries as witnesses, to juries without witnesses with expert witnesses as court advisers, and finally to the system of partisan expert witnesses. No one would now argue against the need to have independent, impartial juries, but there remained some sympathy for the idea of the expert witness as court adviser into the twentieth century, as we shall see later.

Medical experts were the earliest types of expert witnesses, called upon by Coroners' courts to establish cause of death, and they were also important in certifying death in state executions.[44] The earliest case in which scientific expert testimony was used and where the experts testified as partisan witnesses is widely taken to be the 1782 civil case *Folkes v. Chadd*.[45] The issue centred on the reasons for the silting up of Wells Harbour on the Norfolk coast and whether this was due to the building of an embankment. A number of expert witnesses were ranged on both sides, including two Fellows of the Royal Society. The important aspect of this case for the law–science relationship was that the expert witnesses were permitted to express their expert opinions, whether or not they had observed the facts; they were, indeed, 'virtual witnesses'.[46]

Mixed or special juries continued in some form into the eighteenth century and were not formally abolished until 1833, by which time the modern form of trial by jury had been established and the roles of witness and juror had become completely separate, so the expert witness, as we now understand the role, appeared in the eighteenth century and developed in the nineteenth.[47] By the middle of the nineteenth century, the expert witness role was firmly established: 'The battle of the experts had begun.'[48]

The rise of the medical and scientific professions

Identifying the context of early medical and scientific expert witnessing could have the unintended effect of making medicine and science appear more organized into professions earlier than they were because expert witnesses, especially medical witnesses, appeared in the courtroom well before medicine attained the shape of a profession. Indeed, for science and medicine, using a word such as 'profession' can project a cohesion that did not exist onto a loose grouping of activities and people whose occupational experiences and social standing were very diverse. For medicine, Berridge argues that it is necessary to understand how a wide range of medical practitioners worked, including those who were unregulated so that they may be rescued from the 'condescension of posterity'.[49]

Medical practice in Britain would remain unregulated until the middle of the nineteenth century from which time it started to make sense to talk of a medical profession. In terms of the relationship of forensic medicine to medicine as a whole, Watson regards the history of medical practitioners as the most relevant aspect of the history of forensic medicine with physicians, surgeons, apothecaries and midwives all having a role to play.[50] Until the middle of the nineteenth century, medical training was undertaken largely by apprenticeship; only physicians, with their gentlemanly aspirations, obtained a university training; surgeons – think of 'saw-bones' and barber-surgeons – and apothecaries were of inferior status.[51] By the middle of the nineteenth century it was possible to identify medicine as a profession, with consolidated medical knowledge and practice under the control of a medical elite who delineated the boundaries and determined medical registration. The period between 1794 and 1858 has traditionally been seen as the age of medical reform, culminating in the 1858 Medical Act, which was instrumental in setting up a register for all qualified medical practitioners via the General Medical Council, thus delineating legal and institutional boundaries and raising the status of the nascent profession.[52] Importantly, in passing this act, Parliament required that registration was to depend on formal, recognized qualifications.[53] Well before the middle of the nineteenth century, in terms of practice, the old categories of physician/surgeon/apothecary were blurring into the division between a new elite of hospital consultants and general practitioners. Nevertheless, the old distinctions perpetuated as they were institutionalized by the corporate bodies representing the various medical roles. The most prestigious of these were the Royal Colleges of Physicians of London, Dublin and Edinburgh.[54]

The rise in status of medical men in the middle of the nineteenth century was not confined to aspirations towards gentlemanly status amongst the elite of the profession and the exclusion of the unqualified, important though these factors were. There was also the question of how far medical practitioners could define their knowledge as scientific and whether this could be used to enhance their status. At least from the eighteenth century, there was evidence to suggest that doctors regarded their subject as embracing a 'corpus of knowledge which was increasingly defined as scientific'.[55] But it was the 'scientific revolution' in medicine of the middle to late nineteenth century in terms of anaesthetics and antisepsis, the latter eventually finding a scientific basis in bacteriology, which was to consolidate the links between science and medicine.[56] Prominent men of science of mid-Victorian Britain, particularly those who formed the circle of scientific supporters of Darwin's theory of evolution, argued for the primacy of scientific knowledge over religious knowledge as a basis for understanding the natural world. These included John Tyndall (1820–1893) and T. H. Huxley (1825–1895), who advocated the development of a scientific medicine, or at least their view of a scientific medicine, based on the germ theory of disease.[57]

The attitudes of the new medical profession towards scientific attainments and values present a complex picture. It was clear that there was no wholesale titration of scientific theory and innovation into medical knowledge and practice, and it certainly could not be claimed that science made medicine more effective; the rise of the sanitarian movement and its resistance to 'scientific' explanation is testimony to that.[58] On the other hand, it would be wrong to characterize the mid-Victorian medical profession in terms of an overweening concern with gentlemanly accomplishment at the expense of scientific attainment.[59] When the medical world began to see science as representing a form of expert knowledge which could be used to enhance status and emphasize expertise, scientific attainments came to be seen as important. The mantle of expert knowledge with which scientists used to separate themselves from lay knowledge also held appeal in the medical world. Rather than emulating the values of the governors who controlled the hospital system, and therefore much of the professional activity of hospital doctors, the medical profession could use science to emphasize the specialist expert nature of medical knowledge, thereby aiding the separation of their professional activities from lay control. Some of this was a matter of rhetoric. Lawrence argues that in the second half of the nineteenth century, British physicians employed a vocabulary that invoked science as the foundation of medicine, yet prescribed for it only a limited role in clinical practice. Clinical skills were part of the gentlemanly attributes of a physician, and the implication that medicine could be reduced to the rules of an applied science threatened their status as gentlemen. It was a balancing act, and experience had to be seen as the mediator between science and art.[60]

For science, it was not so much a question of regulation heralding the beginnings of professional status; rather, it was opportunity, standing, education and training which were more important. Britain's attitude to its men of science in the early to mid-Victorian periods was notoriously *laissez faire*. In the 1870s a significant investigation into the state of scientific instruction and research was undertaken in the shape of the *Royal Commission on Scientific Instruction and the Advancement of Science*, the so-called 'Devonshire Commission', chaired by Henry Cavendish, seventh Duke of Devonshire.[61] A host of eminent scientific witnesses reported on the parlous state of government support for science, linking this to the future capacity of British industry. The commission produced eight substantial reports between 1872 and 1875, proposing government support for science on a hitherto unheard-of scale, including research grants and the creation of a Ministry of Science and Education. In the event, British liberal values of self-help, fear of centralization and free trade conveniently prevailed, and resulting state support was modest. Had the commission's recommendations been accepted, the professionalization of science and technology would have been considerably hastened.[62] In the event, the foundation of the professional scientific associations was a major building block in scientific professionalization.

Chemists were the main locus of forensic scientific expertise in the decades around the turn of the nineteenth century, and it was the founding of the Institute of Chemistry in 1877 which officially marked the beginning of a profession of chemistry, founded on a 'great boom' in analytical chemistry, including methods centring on metal extraction and manufacture of synthetic dyes and explosives.[63] The essential distinction was between elite research scientists and those whose activities were directed to practical ends, many of whom were analytical chemists, a term which appeared in the London Post Office Directory in 1854, where statesman and chemist Lyon Playfair was one of the first to style himself thus.[64] At least part of the story of scientific status, in the Victorian period at least, rested on the ability of men of science to mount a successful campaign for the legitimacy of scientific knowledge and thereby for the legitimacy of their social positions based on a rhetoric of the ameliorations which science had brought about.[65] Scientific men were just as sensible as medical practitioners to the need to make one's discipline appear useful, effective and progressive, and it was this 'forgotten middle class' who would supply the rising professional class of medical and scientific men in Victorian society with an influence on social thought and the development of industrial society.[66]

From around 1860 the role of the analytical consultant developed based on the application of chemical analytical techniques to industrial concerns, including 'assaying, drug analysis and industrial quality control'.[67] Few companies who needed the services of an analytical chemist could afford a full-time appointment, hence, a range of industries employed analysts for occasional work.[68] Opportunities existed in railways, the gas industry, alkali industry and, of course, in the role of the public analyst whose job it was to analyse food, drink and water for potential adulteration. The Society of Public Analysts, founded in 1874, represented the interests of this group of chemists.[69] Public analysts were, in effect, early forensic scientists, and this was an important route into forensic consultancy.[70]

The early man of science was no less concerned about his status than the medical practitioner, and he was certainly concerned about his salary. By the mid-Victorian era, the age of the dilettante gentleman scientist, itself the preserve of the moneyed upper classes, had long since evaporated. University appointments were still relatively few in number, state appointments were similarly rare, industry posts were developing but not necessarily full-time and none of these posts was well paid. A number of men of science (the term 'scientist' was not widely used until the twentieth century) never held official scientific appointments but funded their scientific careers usually through publishing, lecturing and consultancy – some of the latter was in the law courts. The story of state support for scientific education or, rather, lack of it and the *laissez faire* attitude of the British government through the nineteenth century makes salutary reading.[71]

Class was a major contributor to scientific success. It was very difficult for a man from the lower middle or working class to afford the necessary education to enter a scientific career.[72] In Britain, scientific education was sharply segregated on class lines at all levels. Different educational institutions existed for working-class men.[73] Even scientific knowledge was epistemologically demarcated according to class, with applied science and technical knowledge deemed appropriate for the artisan class – a distinction which persists to this day.[74] A scientifically talented son of a well-to-do family, taking his degree at an ancient university, could expect a very different career path than that of his less-well-off counterpart educated in a mechanics' institute. Hence, it is difficult to overemphasize the importance of class in the making of scientific careers and the separation between the few who could aspire to research careers in the universities[75] and those not of the upper or upper-middle classes who had to settle for careers undertaking routine chemical analyses in state or industry laboratories, such as these existed, or school teaching.

A few scientists of the mid-Victorian era, such as Kelvin with his telegraphic innovations and Perkins with his aniline dyes, could amass large personal fortunes, but this was very far from the norm for the man of science in the Victorian period. Expert witnessing was largely undertaken on a fee basis, even for those with recognized state roles such as public analysts. This state of affairs reflected the opportunities (or lack thereof) for full-time public scientific positions, a state of affairs recognized by the Devonshire Commission but only addressed to any extent by the inauguration of the UK's Department of Scientific and Industrial Research (DSIR) in the twentieth century, when it became clear that Germany had command of much more efficient war technologies in the First World War.[76] The DSIR was set up as a government department with administrators and an advisory committee to encourage scientific research for industry.[77] The National Physical Laboratory and Government Chemist's Laboratory would come under its ambit in due course. So models for careers as scientific civil servants already existed by the 1930s when full-time, salaried state positions for forensic scientists became available in England. However, the new forensic science positions were, of course, part of the Home Office and not the DSIR and were strongly linked to police organization. Even then, many expert witnesses continued to earn their living through independent consultancy. Indeed, as Chapter 4 outlines, there were senior voices in the police who preferred a system of independent experts, as it gave police forces a free hand in whom they chose to employ for expert analyses.[78] Given that so many career options did not attract a full-time salary, whatever the difficulties surrounding courtroom appearances, the option of earning fees from expert witnessing was as attractive to many scientists and engineers as it was to medical men.

Good science – bad science: claiming authority in the courtroom

For all its problems, the courtroom was a major locus where scientists could claim or display their authority, a place where new scientific techniques could be accepted or rejected; hence, it was a place where scientific knowledge was made. In their courtroom appearances, we can see the strategies that scientists adopted to 'perform' scientific authority. In addition, the workings of scientific expert witnessing in the late nineteenth and early twentieth centuries reveal interesting aspects of the epistemological status of various forensic sciences and techniques and how these were granted epistemic authority. Saks argues that there was a difference between forensic techniques developed specifically for the law and techniques based on existing science, usually chemical analyses.[79] The former were mainly identification sciences and technologies such as fingerprinting, identification of marks and handwriting analysis – in other words, the technologies of individual identification, many of which were to come under the banner of 'police technique' in Britain by the middle of the twentieth century. The most important of these, fingerprinting, was accepted as unproblematic by the 1920s and 1930s, the period when forensic science was developing in the twentieth century in the UK. So when the first of the *Home Office Forensic Science Circulars* (1936) debated new scientific methods of capturing and preserving fingerprints based on research in the UK and on the continent in considerable detail, it had no need to discuss whether fingerprints were unique to the individual or whether trained fingerprint officials could reliably match fingerprints; these things were not regarded as problematic.[80] Of course, the reliability of some of these forensic techniques was a scientific pigeon that would come home to roost much later in the twentieth century when scrutiny of the scientific bases of several techniques would come under the microscope, but they were not in the spotlight in the early and middle years of the twentieth century.[81]

Agreeing with Saks, Caudill makes a distinction between 'normal forensic science', which broadly refers to chemical analyses, and, on the other hand, forensic identification science and sees the problem lying mainly with the latter, as these techniques are not underpinned by basic science.[82] Saks is especially harsh on identification technologies.

> The forensic identification sciences have no basic science to undergird them. For most of their history, the forensic identification sciences had little or no academic or industrial infrastructure to provide them with knowledge, resources, or personnel. Instead they invented themselves and they exist on their own . . . consisting of nearly all application and no science.[83]

This is a complex story, largely outwith the scope of the present work, as discussion centres on a more recent period. Nevertheless, the debate rests on the assumed objective nature of science to produce analyses which are more reliable than those not based on 'basic science'. This is one of the key issues in the National Academy of Sciences's recent report criticizing forensic science provision in the USA.[84]

However, we cannot assume that techniques which are *not* based on an idealized concept of scientific method will necessarily become viewed as unscientific and therefore be seen epistemologically problematic at some stage in their careers.[85] In keeping with recent scholarship in science and technology studies, it is a question of avoiding a presentist conception of 'correct' and 'erroneous' science. In any case, despite specialization in the forensic sciences, forensic scientists of the twentieth century had to develop their specialist skills alongside the skills of finding and using whatever technique suited the purpose, as the words of a British forensic scientist from the 1980s attest in a plea for forensic scientists to develop broad skills alongside specialist skills. 'There's no such thing really as forensic chemistry and there's no such thing as forensic biology. You're a plagiarist and you pinch whatever technique is available to meet the demands of the situation you are dealing with.'[86]

Mnookin describes handwriting evidence as a paradigm example of a scientific technique which was specifically developed for the law and which, despite the ambivalence of members of the legal system, became accepted as an appropriate form of proof where:

> Handwriting experts, initially considered to be the lowest possible order of proof, greatly improved their image and increased their credibility by selling the judges something they wanted: explanations about authorship that looked reasoned, credible, and scientific, and that the jury could (at least seemingly) understand.[87]

In potential contradiction to Caudill and Saks, Mnookin argues that the means by which handwriting experts were able to demonstrate their authority in the courtroom was significant. The two main ways for a scientific witness to perform authority were by achieving courtroom respect for an accepted scientific or medical technique, especially when undertaken by an acknowledged eminent representative of the discipline, or where the technique and individual did not have the kind of status which warranted deference, instead they tried to achieve authority by educating the judge and jury.[88] Handwriting experts were not unique in using techniques of demonstration and education for their evidence, but they were different in that education was the main way they presented their evidence as legitimate.[89] The courtroom, therefore, became a crucible for legitimating knowledge.[90] This was especially important for experts who could not lay claim to accepted authority for their discipline. Visual presentation was particularly important in persuading judge and jury, and photographs were especially effective as they appeared to offer objective evidence. Visual demonstrations did not necessarily mean that a technique was reliable, rather that they were 'often effective as rhetorical displays'.[91]

In the late Victorian period, the courtroom was an important site for the application of science, and appropriately qualified scientists were much in demand as experts on technical matters in relation to gas, water, electrification, patents, pollution and adulteration, insurance, and, of course, in relation to criminal cases.[92] Expert witnessing was as important in civil cases as in criminal cases. For instance, patent disputes supplied a major source of expert witnessing opportunities.

> The significance of the legal system . . . is that many decisions about who had the right to profit from novel electrotechnology, and to be recognized as its "true" inventor, were made in the courtroom by a judge weighing rival claims of inventive history . . . they were resolved by the ordeal of the adversarial system. This approach pitted defendants directly against plaintiffs in open semipublic contests in which opposed patentees, company secretaries, and their hired expert witnesses competed with each other as authorities to win a judge's assent to their claims.[93]

However, many scientific expert witnesses had a difficult entry into the legal arena. As Golan notes, it was a free marketplace where clients 'shopped around' the various sources of available expert advice. The courtroom was a far call from the lecture theatre or laboratory. Here, they were shut into the witness box:

> . . . excluded from the legal decision-making processes, and manipulated as mere tools in the hands of the lawyers. Browbeaten and set against each other by the lawyers, scientific witnesses quickly found that their standard strategies for generating credibility and agreement did not well withstand the adversarial heat of the courtroom.[94]

It is hardly surprising, then, that the system of giving scientific expert testimony, whether in civil or in criminal cases, in the middle to late Victorian period was widely taken in the press, popular culture and in legal circles to be highly problematic.[95] Some believed that the problem was created by an adversarial legal system that set scientists called for prosecution and defence against each other; such a view implied that there would be a scientific solution if it were not for the system, and this agrees broadly with the findings of Smith's study.[96] However, a sophisticated level of debate and awareness of the matter was evident in the late Victorian period in the thinking of certain scientists, an approach which acknowledged that there was no simple 'right answer' and no simple solution.[97] Ranged on one side was William Odling, President of the Institute of Chemistry, Professor of Chemistry at Guy's Hospital and the Royal Institution, who argued that scientists needed to make a living, consultancy work was as important as research and the two were not incompatible.[98] Odling's argument was directed at the scientific elite who tended to look down on practical scientists – the very people who were imaginatively applying scientific methods to the wide range of problems which were arising in newly industrialized Britain and who were ignored by the great and the good.[99] On the other side, Norman Lockyer, elite astronomer and editor of *Nature*, the premier scientific periodical, spoke up for the purity of scientific research, arguing that expert witnessing was 'undignified' for scientists.[100] Although the Lockyer–Odling debate was about the use of scientific expert witnesses in court, it was also a debate about the profession of science in Victorian Britain and the ethical duties of men of science in relation to their profession.[101] This was a significant moment in the history of scientific professionalization.

Although courtroom witnessing work was an important source of income for many men of science, such work could make them appear to be 'hired guns', tailoring their evidence to fit the case of the party that had paid them.[102] Further criticisms, which remained important for the forensic scientists of the twentieth century, were, given that expert witnesses were able to give opinions based on their scientific judgment (as opposed to matters of fact, which was the province of the ordinary witness), how could juries make a decision when that scientific opinion disagreed with other scientific opinion? A further problem was that scientific expert witnesses did not necessarily prepare appropriately or put their evidence across effectively. A scientist who was otherwise authoritative just might not be able to put in a good performance as a courtroom witness. Indeed, skilled lawyers often deliberately ran circles round witnesses who were delivering their evidence in good faith.[103] Even a man of science as worthy as Michael Faraday had his evidence demolished in court in an insurance trial early in his career.[104] It was small wonder, as we shall see later in the chapter, that the forensic scientists and pathologists of the 1920 and 1930s were wary of the courtroom and developed their own strategies for coping with the experience.

Warring scientists in the courtroom – the Palmer case

As an example of the long historical reach and problematic nature of expert witnessing and 'public scandals of scientific expert testimony',[105] we turn to the notorious Palmer case as a prime example of the problem of scientific expert witnesses. The Dr Palmer strychnine poisoning case, 'the crime of the age',[106] provides an exemplar, illustrating a number of the principal reasons why scientific witnessing in legal trials was seen as such a problem, and it has rightly received considerable attention.[107] In 1856, Dr William Palmer was charged with murdering his friend, John Cook, with strychnine. It is likely that he murdered several other people, including his wife and brother. An inveterate gambler, Palmer's motivation was money. Golan suggests:

> The fact that such atrocious wickedness was consistent with good breeding and education attracted unprecedented public attention to his trial. The trial was considered by the legal historian, Sir James Stephen to be one of the greatest trials in the history of English law.[108]

Given the high-profile nature of the case, the prosecution and defence teams' wrangling over the detection of strychnine in the post-mortem was highly visible and had all the more potential to cast an unfavourable light on the medical and scientific experts involved. Palmer's was the first strychnine poisoning trial in the UK. An incredible thirty-nine expert witnesses offered evidence. These included Dr Alfred Swaine Taylor, FRS, Professor of Chemistry at Guy's Hospital, the leading forensic medical expert of the day, he who had produced the canonical work of medical jurisprudence which would still be in print a hundred years later.[109] Acting for the prosecution, Taylor was one of the medical men who had undertaken the post-mortem. He failed to find strychnine, but he theorized that it had been absorbed into the body and chemical changes meant that it could not be detected. However, the defence argued that Taylor's inability to find strychnine proved it was not present.

William Herapeth, professor of chemistry at Bristol Medical School, rejected Taylor's theory, which he saw as an invention for the trial, arguing that if strychnine had been present, Taylor would have found it. Public opinion was already hostile to the contradictory scientific evidence of the trial when Herapeth, acting for the defence, confessed on cross-examination 'that he had again and again bragged among friends that he too thought that Cook was poisoned by strychnine but that Taylor did not know how to find it'.[110] The scientific and expert witnesses involved in the trial could hardly have looked worse as they appeared to be guilty of incompetence, inventing theories and acting in a partisan way. Small wonder that the attorney general, Sir Alexander Cockburn, and Chief Justice Lord Campbell, the trial judge, roundly criticized the scientific witnesses, especially the behaviour of the defence. 'The newspapers that eagerly followed every detail of the trial quoted this unprecedented rhetoric of the attorney general and chief justice widely. Palmer was convicted and with him much of the public image of the scientific community.'[111]

The Palmer case was highly complex, and my intention is not to attempt anything like a full account.[112] Nevertheless, even a brief outline of the case readily illustrates some of the difficulties involved when experts were lined up against one another. As Watson notes, the case could not have been decided purely by chemical analysis; instead, the medical, circumstantial and scientific evidence had to be evaluated by the appropriate cross-examination of medical and scientific witnesses.[113] Salient points for the current discussion centre on the level of disagreement and the personal failings manifest in the behaviour of the medical and scientific expert witnesses. The adversarial system in English and US courts naturally encourages disagreement, and expert witnesses acting for defence and prosecution will, of necessity, disagree. In a subject which looks for consensus on scientific facts and where several centuries of scientific achievement had contributed to the view that the 'man of science' was a dispassionate observer of facts, cases such as the Palmer trial showed scientific proceedings in a poor light, reinforcing the view, if any such reinforcement was needed, that the adversarial system, which potentially forced scientists into disagreement, could be seen as bad for science.[114]

Expert witnessing in the twentieth century

These problems did not disappear in the twentieth century. The Frye standard for scientific evidence, adopted in the USA in the 1920s, and, in the 1990s, the Daubert test, were, in some part, efforts to mitigate the problems of expert witnessing at least in terms of admissibility of scientific evidence and only in the USA.[115] Later in the twentieth century, the term 'junk science' was popularized by Huber, more so in North America than in the UK, to describe the deliberate attempt to peddle fraudulent science in the courtroom. However exaggerated Huber's account may be, there is plenty of contemporary criticism of standards of scientific evidence in the criminal courts.[116]

Although there are burgeoning literatures on the problem of scientific expert witnesses in court in Victorian times and in present times, there is relatively little scholarship and contemporary material which claims that there was an expert witnessing problem of the same magnitude in the 1920s and 1930s. Part of the reason appears to lie in the fact that there were by then a number of established scientific expert roles in the UK. In the medical arena, these included Home Office Analysts and pathologists and public analysts for food and substance adulteration, and there were several well-respected, independent scientific and medical expert witnesses, chief among these being the incredibly authoritative Bernard Spilsbury, whose influence is difficult to overestimate. A considerable boost to forensic authority was given by the widespread acceptance of identification techniques such as fingerprinting. In other words, the forensic identification problem of the individual criminal which beset the latter half of the nineteenth century and earlier part of the twentieth century appeared, for the time at least, to be solved (no matter what was to come by way of critique later in the century). In addition, there was the promise that analytical and optical scientific techniques developed to deal with trace evidence would make significant inroads into crimes against property – a view promulgated by the 'scientific aids' movement of the 1930s and the establishment of forensic laboratories in England and Wales as Chapter 4 describes. Although there were detractors from the enthusiasm to set up a system of state forensic laboratories, the new laboratory techniques and organization offered optimistic prospects that science could be used effectively in the fight against crime. All these factors tended to mitigate against some of the outright wars of words prevalent in earlier trials. Nevertheless, scientists knew that battles in court reflected badly on them, and part of the rationale for developing appropriate strategies to deal with cross-examination could also be seen as the need to keep a lid on the potential for major courtroom disagreements to erupt.

In any case, the spotlight was falling onto policing and scientific detection as the weakest links in the story of forensic analysis rather than presentation in court. The new Home Office forensic laboratories, in their formal position in relation to the criminal justice system, played a role in promoting 'prosecution mindedness'.[117] This contributed to the prosecution gaining the upper hand forensically, tending to militate against dissenting experts in court. However, there was an ever-present danger that a scientist working closely with the police would accept a police hunch and read his scientific results through this, a problem which was recognized by some forensic scientists.[118] For instance, Julius Grant's (1901–1991) *Science for the Prosecution* was a popular forensic science book in the 1940s.[119] There was no equivalent book called *Science for the Defence*, a point which was made in a review of the book by an independent scientist who appeared in prosecution and defence cases:

> [T]he title he has chosen may be misconstrued. It tends to imply that the resources of science are the monopoly of the prosecution. The possibility of such misconception could be removed by adding two words to the title – "Science for Prosecution *and* Defence."[120]

However, the problem of science in the courtroom certainly did not disappear in the first half of the twentieth century, and in the UK, the words of scientific witnesses themselves provide the means of uncovering some of the perceived issues. Charles Ainsworth Mitchell (1867–1948), the reviewer of Grant's book, was an analytical chemist and author, an acknowledged leader in the development of the new forensic science in the twentieth century and was much in demand as an expert witness in document cases in the first half of the twentieth century. Mitchell produced three popular books on science and the law.[121] Although the titles would suggest that the reader could expect much detail of the business of expert witnessing in the first half of the twentieth century (*Science and the Criminal* (1911); *The Scientific Detective and the Expert Witness* (1931), *A Scientist in the Criminal Court* (1945)), in common with many of his contemporaries' autobiographies and semi-popular books on forensic matters, understandably the emphasis is on interesting cases which display the use of science in criminal justice, so clues on his views of the witness role must be mined from the text.

In the earliest of his popular books, Mitchell's rhetoric centred on a reiteration that scientific evidence is objective and impartial and the belief that defendants should have access to scientific expertise in the same way that they had a right to legal representation even if they could not afford it.

> This principle might well be extended so as to cover the ground of scientific evidence. Under the present conditions the prosecution has unlimited facilities for applying every description of test, but it has not always been easy for the representatives of the accused person to obtain scientific help in criticising the nature of this evidence.[122]

As an independent consultant, Mitchell was able to act for prosecution or defence and was therefore better able than the Home Office forensic scientists who were employed for the prosecution, to see the disadvantages that the defence might labour under. He could see the advantages of making scientific analyses available to the defence. Despite the declared impartiality of scientific evidence, if only the prosecution had access to scientific expert witnesses, then a point that could have been found in favour of the accused might well be overlooked.[123] For example, suggested Mitchell, suppose a chemist for the prosecution found that a stain on the accused's clothes was blood, then that would go against the accused even though the expert witness could not say whether it was human blood or not.[124] But let us assume that the stain was due to rabbit blood, and the scientific expert witness for the defence knew about what were at the time relatively recent methods of distinguishing the blood of different animals and was able to determine the origin of the blood stain. This would 'break one of the links in the chain of evidence'.[125]

Rather than seeing multiple experts as a source of conflict which reflected badly on scientific expertise in the courtroom, Mitchell regarded the use of multiple experts as positive, likely to mitigate against mistakes and potentially fairer to defendants. The opportunity to examine scientific evidence first-hand rather than just criticizing the results of scientific analyses was a potential advantage of employing experts for prosecution and defence. If the defence counsel cross-examined a prosecution expert witness, the lawyer would be unlikely to have the specialist knowledge to point out weaknesses in the scientific evidence; this is why expert witnesses were required on both sides. Mitchell made an important point that, in this respect at least, a poor person on trial could not afford to get scientific support and so was much disadvantaged with respect to a rich person who could afford to get the best scientific advice.[126] Mitchell was apparently arguing for access to scientific evidence as a means for furthering social equality, but, at the same time, was making it clear to his readers that he regarded science as inherently impartial and of considerable utility in criminal cases. As the years passed, defendants, of course, became much more knowledgeable about the ways in which forensic evidence could be used against them and the need to have the support of their own scientific expert witnesses.[127] An example from the 1960s, where the defendant in a murder case retained a scientist and a gunsmith to act on his behalf, makes this point. In his statement to the court, the defendant gave the reason: 'he had been "forensicated" on a number of previous occasions and he now wished to call his own expert evidence'.[128]

Mitchell was an expert on inks and forged documents, often working on cases involving disputed wills. In his final popular science and crime book, he reported the unusual way in which he was instructed in a particular trial; this was, he felt, a much better way of using an expert witness.[129] He was given the disputed will for examination and a bottle of ink, which was said to have been made from a mixture of three inks, from the house where the will was alleged to have been signed. He was asked not to communicate with either side. He found that the constituents of this unusually mixed bottle of ink were such that they explained the peculiarities of the ink of the signature. Mitchell gave this explanation to the court in the witness box, and the judge ruled that the signature was genuine and so found in favour of the deceased man's wife who had brought the case to court. Mitchell was not cross-examined.

This was not in itself a significant case in terms of the scientific evidence, but it was significant in that Mitchell was put in the position of acting as a court adviser rather than being called for the plaintiff or defence. More than thirty years after he first expressed a similar idea in print, Mitchell reiterated his support for courts making the selection of witnesses, given that in the adversarial system, each side selects its own experts, who may not, in any case, be properly qualified. He argued that having a panel of witnesses from which one could be selected would ameliorate the situation. Other alternatives included having the judge question witnesses on both sides or the judge directing expert witnesses to work on a technical point and produce a joint report.[130] The latter reinforced Smith's point that expert witnesses believed they could reach some measure of agreement as to opinion if they were able to discuss their evidence outwith the confines of the adversarial system.[131]

It is notable that Mitchell made no reference either to the work of the Metropolitan Police Laboratory, which had been in operation (although not entirely successfully) for the better part of a decade when his book was published, or to the work of the regional Forensic Science Laboratories. His experience was of an earlier period of expert witnessing, an 'old guard' style of expert witness belonging to the era when expert witnessing in court was firmly in the hands of individuals. Nevertheless, he was promoting a thoughtful, fairer and less adversarial view of the way in which expert witnesses could be used in English cases.

Professor of Forensic Medicine Sydney Smith (1883–1969) also argued that a scientific expert witness should adopt the same attitude whether he was called for defence or prosecution and should be ready to make his evidence available to the defence whether or not the defence was able to pay a fee. Like Mitchell, he pointed out that the Crown had vast resources at its disposal, including the whole of the police and Criminal Investigation Department (CID) apparatus and specialist laboratories. The defence was likely to be short of resources and money and was, therefore, usually at the mercy of the critical appraisal of prosecution evidence in the courtroom.[132] Smith was often asked why he had given evidence for the defence, given that the accused's life might not be worth saving. His reply was: 'the life of the accused may not be worth saving, but the principles of justice always are'.[133]

Spilsbury in the witness box

In terms of managing performance, the grand master of the witness box was the Home Office forensic pathologist Bernard Spilsbury (1877–1947), who rapidly rose to fame as a 'celebrity pathologist' after his stunning performance in the Crippen murder case, a case which has many interesting aspects for the history of forensic science and which is discussed in some detail in Chapter 3.[134] Spilsbury was well known for his extraordinarily authoritative courtroom performances, unshakeable self-belief and influence on juries.[135] He gave evidence in nearly forty years of murder trials, usually for the prosecution, often worked alone, was obstinate and unassailable, giving professional opinions on matters well beyond his professional expertise. Unfortunately, it appears that he was wrong on several occasions, and a number of convictions based on his evidence would now be regarded as unsafe; it is very likely that he sent some innocent people to the gallows.[136]

Posterity has not been kind to Spilsbury; colleagues admired him but did not like him and, after his death, were highly critical of his approach.[137] However, he displayed a number of important qualities which enhanced his performance in court. Spilsbury could put across complicated medical evidence in ways palatable to a jury. Mnookin suggests that education and demonstration often had to substitute for authority in expert witnessing.[138] Spilsbury was a master of both: 'a brilliant cross-examinee'. He 'exemplified a technique likened to a neat drop shot at Wimbledon, giving bland answers in cross-examination until, quite unexpectedly, he would ram home his original finding'.[139] He looked the part as well as acting the part. Chapter 5 discusses self-image of a number of forensic scientists through the conduit of autobiographies and biographies, noting the care that such works take in presenting an appropriately scientific image of the scientist in the laboratory or crime scene. Spilsbury's aristocratic bearing contributed to his growing media profile. His success in court in connection with high-profile murders cases and his ability to be completely definite about his professional opinion put him in the centre of 'a prosecution culture used to monopolising the process of forensic fact-making'.[140] However, Spilsbury's published output was meagre; he never produced a textbook or autobiography and was not given to the kind of reflection on law and science in print that some of his forensic contemporaries undertook, so we have few clues as to his self-awareness in terms of his courtroom performance; nevertheless, as Rose claims, he 'shrewdly manipulated the media and dominated entire trials'.[141] He was clearly fully aware of his devastating effectiveness.

Indeed, Spilsbury's domination of forensic pathology and his style of expert witnessing has had a lasting influence, extending at least into the 1980s: 'still acting as a yardstick against which the forensic pathologist is judged'.[142] Green argues that Spilsbury's far-reaching impact was largely negative and has contributed to an unhelpful stereotyped view of forensic medicine which the discipline struggled to shake off. Indeed, the very qualities which made Spilsbury such a successful performer in the witness box continued to haunt the discipline nearly fifty years after his death. Prominent amongst these was the idea, widely held in criminal justice authorities, that forensic pathology offers absolute proof as an 'exact science' and that 'any "expert" worth his salt' would go into the witness box and undermine a counter argument by insisting his view is correct and by using rhetoric rather than by making a valid analysis of the evidence.[143] Furthermore, Green argues, Spilsbury's style of work has contributed to the idea that it is possible to give an immediate diagnosis at the scene of a crime and a microscope is not required. Understandably, this promotes a culture where everyone involved in court proceedings expects too much.[144]

Although Spilsbury achieved considerable forensic authority in the courtroom on his own, his performance was considerably enhanced by the presence of gunmaker and ballistics expert Robert Churchill (1886–1958) in several shooting cases. '[T]he pair would perform together like clockwork, a smooth double act.'[145] Churchill had some difficulty building a working comparison microscope for comparison of bullets; he was not, after all, a scientist.[146] But after visiting Calvin Goddard, the prominent US ballistics expert, at the Bureau of Forensic Ballistics in New York in 1927 and having seen a comparison microscope in action, he came back to have one built for his own use.[147] Churchill quickly became seen by the Home Office as their preferred firearms expert in the 1920s. His biography paints a picture of a formidable character, involved in almost every shooting case in England for a period of nearly fifty years, becoming a celebrity, on the front page of newspapers when he was called as a witness in a shooting trial and revelling in the 'glamour' of murder cases; later in his career, someone who missed the 'old days' with the ever-present threat of the hangman's noose.[148] As his biographer notes: 'Men hanged on what he saw through his microscope'.[149]

Spilsbury and Churchill first worked together for the prosecution in a shooting case in 1913.[150] The remarkable point about this and later shooting cases where the two were teamed was that Spilsbury, a pathologist, undertook experiments, firing guns into appropriate materials at different measured ranges. It seems incredible now that ballistics evidence could be given in court by a pathologist and that evidence could be admissible, but he was not the only medical expert to be shooting guns and using the results in support of evidence in court. Possibly the most notorious case where the formidable team of Spilsbury and Churchill gave evidence was the Merrett case of 1926 in Edinburgh.[151] This case was unusual for Spilsbury in that it was north of the border and he appeared for the defence.

While sitting at her table, Bertha Merrett had been shot, the police were called and, on her son's evidence, treated the case as one of attempted suicide. The scene was tidied up, and much forensic evidence was lost or overlooked. The question of whether or not there was blackening round the bullet wound was a potentially crucial piece of evidence, as the presence of blackening would have indicated shooting at very close range, making the suicide explanation more probable. She was taken to hospital, treated as a mental patient – and therefore not properly questioned – and she died after lingering for a few days.[152]

Harvey Littlejohn (1862–1927), Professor of Forensic Medicine at Edinburgh, acting for the prosecution, where the son, Donald Merrett, was the accused, undertook some shooting experiments with his opposite number from Glasgow, John Glaister Jnr (1892–1971), using skin from an amputated leg at the suggestion of his old pupil, Sydney Smith. Smith was, at the time, the Medico-Legal Expert in Egypt but would eventually become Littlejohn's successor in the Chair of Forensic Medicine at Edinburgh. Smith happened to be on holiday in Edinburgh that summer and was able to witness the progress of the trial.[153] Originally believing suicide was a possibility, Littlejohn subsequently submitted a report stating that 'suicide was in the highest degree improbable' because there was no blackening of skin round the wound which was the expectation for a gun fired at close range.[154] Churchill and Spilsbury were hired for the defence: 'formidable, expensive and determined'.[155] Smith described Spilsbury, whose path he would cross several times in murder cases over the succeeding years, as 'very brilliant and very famous, but fallible like the rest of us and very, very obstinate'.[156] Churchill was also stubborn and dogmatic. '[T]hey were indeed a formidable team terrifying when they made a mistake, as they did here.'[157]

Spilsbury was determined that the defence should make its own shooting experiments, so he and Churchill experimented with shooting an amputated leg.[158] Churchill did not think the experiments were useful, as dead flesh does not behave the same way as live flesh.[159] Later, Spilsbury and Churchill were able to experiment with the actual gun from the crime scene, but they did not use exactly the same ammunition. Smith regarded it as 'defence by confusion'.[160] The trial continued with Spilsbury and Churchill, as it were, sticking to their guns and with the defence lawyer at one point accidentally calling Spilsbury 'Saint Bernard'.[161] The jury returned a verdict of 'not proven'. Merrett was acquitted, only to murder his wife and mother-in-law in 1954.[162]

'A clash of scientific evidence always makes a jury skeptical about the whole scientific evidence, and it was especially important here because of the poor quality of the police evidence.'[163] The bungling of the police 'and the credit given to the misleading evidence of Spilsbury and Churchill, who had made a mistake and were too stubborn to admit it, allowed Merrett to live'.[164]

Smith was to encounter Spilsbury again in a 1929 murder case where Spilsbury had undertaken the post-mortem and Smith was acting for the defence, by now installed in the Chair of Forensic Medicine at the University of Edinburgh. In this case, Smith did not believe that Spilsbury's account of death was correct, allegedly by manual strangulation as indicated by a bruise on the larynx, the latter only observed by Spilsbury and no longer visible.

This was his experience of the hostile cross-examination that followed. The barrister was sharp and sarcastic, insisting that Smith should answer the question even though it might be difficult for 'a gentleman who lectures' to answer questions.[165] The barrister tried to manoeuvre Smith and the other witness for the defence into a position where they could be seen to be belittling Spilsbury, another tactic in cross-examination of expert witnesses. At one point the prosecution counsel read out a passage from Taylor's *Medical Jurisprudence* on the topic of asphyxia, asking if Smith agreed with it.[166] As Smith was the current editor, he could not very well disagree; fortunately, it was not in conflict with his evidence. As already noted, quoting the scientist's own textbook to him was a potential strategy for counsel to outflank an expert witness.

A similar example can be found in relation to John Glaister Jnr while he was still a fairly junior lecturer in medical jurisprudence and his father was Professor of Forensic Medicine at the University of Glasgow.[167] Glaister Jnr had recently completed his MD degree which involved research on the precipitin test to establish whether a source of blood was human or animal; he had also examined bloodstains on a range of materials. In the case in hand, he was to give evidence based on his research.[168] Both Glaisters were appearing as witnesses for the prosecution. The defence lawyer questioned Glaister senior by asking whether he had said in his textbook that such tests for blood were at an early stage and were not fully reliable. Glaister senior responded that, having seen further work, his views had changed, and his new opinion was to be incorporated in the next edition. This gave the defence advocate the opportunity to ask, given that he had already changed his mind, how were the judge and jury supposed to know whether or not he would have to change his mind again?

Glaister Jnr appeared in the witness box next, having been outside the court while his father was cross-examined and therefore unaware of what his father had said about his change of opinion on tests for blood. The defence asked Glaister Jnr a series of questions that appeared to probe his knowledge of the subject. But he began to think that the defence counsel was deliberately setting a trap to establish the son as the expert and to undermine the father to imply that the latter knew little about the subject and was merely relying on his son's judgment. Realizing that he was being set up, Glaister Jnr stated that he had done all the tests himself but they were checked and agreed by his father. 'The trap had come close to springing – but it had been avoided.'[169]

Hamish Walls and 'witnessboxmanship'

No other forensic pathologists and, indeed, none of the forensic scientists of the first half of the twentieth century who form the focus of this book, achieved anything like the celebrity status of Spilsbury. However, some were prepared to reflect on the business of the performance of the expert witness in the courtroom, drawing from personal experience, and these reflections are often illuminating, particularly in relation to the achievement and maintenance of scientific authority and the presentation of facts and probability.

James Brierley Firth (1888–1966), the first Director of the North Western Forensic Laboratory, certainly believed that the considerable experience he had built up as a scientific consultant enforcing the River Pollution Act for the Ministry of Agriculture and Fisheries was a major positive quality when he was being considered for the forensic laboratory role. Intelligibility was a major part of the expert witness's courtroom performance.

> [N]ot so much what you know but *how* you put it over; one must be able to deal with the matter under review in such a way that anyone can understand it . . . The ability to put scientific evidence before a court in a form intelligible to juries and others has been extremely useful to me.[170]

The most forthcoming forensic scientist on expert witness performance in the courtroom was Hamish Walls (1908–1988). Walls spent some thirty years as a forensic scientist, starting out in the 1930s in the Metropolitan Police Laboratory, before moving to Bristol, then Newcastle and then back to the Metropolitan Laboratory as Director before retiring in 1968. Chapter 10 of his autobiography makes it abundantly clear how he felt about the courtroom side of his role.[171] He was forthright about the tricks that counsel could play and how the expert witness could respond. In his chapter 'Uneasy bedfellows: Science and the law', Walls suggested that forensic science differs from other applications of scientific laboratory techniques, in that it appeared to exist 'to put the scientist into the witness box to be shot at'.[172]

For Walls, the witness box was a 'lonely place' for no matter how much help you got from other colleagues, in the witness box you were effectively on your own.[173] He argued that the different ways that scientists and lawyers think often led to misunderstandings. He found it surprising how some lawyers seemed to believe that science was too difficult for them to understand. Walls recounted one drink-driving case where in his evidence he gave the standard formula of multiplying the urine alcohol level by three quarters to arrive at an estimate of the blood alcohol level.[174] The lawyers claimed to find this far from simple, although Walls acknowledged that this may have been a strategy to persuade the jury that the evidence was too difficult for them to understand.

When Walls started out in the 1930s, legal practitioners still regarded science as a fixed and established body of knowledge whose main business is a kind of Baconian fact collecting.[175] Small wonder, then, that the definitive way in which

Spilsbury presented his evidence should be so appealing to the law courts. Walls hailed from a later generation of forensic scientists than scientists such as Mitchell; indeed, he was forty years Mitchell's junior. Forensic scientists of the later era, who spent most of their careers in state forensic science laboratories working as part of a team, were able to depend more on that badge of office rather than on personal authority. Individual celebrity status was not necessarily required to reinforce the authority of their testimony. Wider public knowledge of science, coupled with its somewhat diminished authority suffered in the wake of World War I, meant that it was more difficult to rely on widespread acceptance of the objective, fixed nature of scientific evidence.[176] Other strategies to convey less-than-certain knowledge to the courts were needed.

An important factor illustrating the problems of communication between scientist and lawyer relates to the precision with which a measurement can be made. All scientists know to expect experimental errors, and as long as they know how large the error can be or what the mathematical probability is that the real value differs from the measured one by a known amount, they have no problem working with this; it is all part of the business of doing science. However, the indefinite quality of probabilistic reasoning was not, in Walls's view, very palatable to the judiciary in a legal trial. Trying to put this idea over in court was liable to produce a response where the judge asks for facts, not probabilities, and the scientist is treated as if they have done something shocking by putting probabilities before the court in the first place.[177] Uncertain knowledge had to be presented obliquely. 'In my experience you don't mention probability; it's a dirty word. Like shouting a four-letter word in church or something. If you said the odds of so and so it was accepted.'[178]

Walls's successor in the role of Director of the Metropolitan Laboratory confirmed that lawyers found probabilistic explanations unpalatable, so the expert witness had to have an explanation strategy. For instance, rather than using the expression 'one in a million' to the jurors, you might refer to fifty-five people in Britain; this was the same ratio, but more tangible and easily understood. The trick was to present the truth in a different form or make it more understandable, such as explaining infra-red spectroscopy and atoms vibrating by using the analogy of tuning a radio, where one position on the dial is the frequency of the station you want.[179]

However, Walls recognized that the job of the scientist and the lawyer are more similar than the lawyer often realizes. It was relatively easy to frame legal and scientific reasoning in similar hypothetico-deductive ways. The scientist sought to see whether the consequences of his hypotheses agree with experimental data; similarly, in a criminal trial, the prosecution would effectively hypothesize that the accused person is guilty, supplying data to support this hypothesis, i.e., evidence.[180] However, the data which the law is obliged to accept are, by scientific standards, incomplete. The law cannot set up control experiments and cannot re-check experimental results. And, of course, whereas science can use all the relevant data it collects, there are strict rules about what evidence can be used in court.[181]

Neither science nor law can offer complete certainty, but where a scientist would like this to be expressed mathematically, the law has 'beyond reasonable doubt' which tends to be defined circularly.[182] Despite the way that some seemed to expect that this would all be handed over one day to 'Mr Justice Computer', Walls knew that it simply was not possible to give a number to those things which judge and jury have to weigh up in deciding guilt or innocence.[183] In this respect, Walls was displaying a level of sensitivity to the limits of scientific evidence. For instance, he gave the example of a hypothetical case where, if 1 person in 10,000, say, had a particular blood type and bloodstains on clothing from the scene of the crime were of that type, then it is easy to give a reliable figure for the probability that the blood belonged to the accused. However, suppose you had a situation where the blood group was much less rare, so any one of many thousands of people could have matched. Imagine that there was evidence that the crime was likely to have been committed by someone local and the accused was local. A jury was likely to set great store by the latter fact, yet you could not put a numeric probability to it in the way that you could for the probability of blood group.[184]

The differences in legal and scientific proof and the problems of communication between lawyer and scientist were just as problematic for medical witnesses. Except for experienced pathologists and police surgeons, Walls did not believe that doctors made good witnesses.[185] But this is understandable, as all their training led them to draw tentative conclusions from inconclusive biological data. Therefore, doctors were often unable to give the kind of certain answers which law courts would like. Of course, witnesses had to be truthful, but they also had to present the truth in such a way as to appear appropriately authoritative. 'The courts want to hear authority issue from the lips of their expert witnesses.'[186]

As it was all too easy for an expert witness to look foolish if he was wrong-footed by having a technical question which he could not answer fired at him at the beginning of the cross-examination, Walls had developed an effective strategy for dealing with the perils of the courtroom.[187] Roger Smith noted that, in his study, forensic experts often used games language for 'reconciling an adversarial strategy with individual integrity, or the institutionalized commitment to winning a case with the personal skills of describing scientific proof'.[188] Indeed, Walls himself ascribed to the view that the English were wedded to the adversarial system because of their fondness for cricket to his old boss at the Metropolitan Laboratory, L. C. Nickolls: 'You put in one side and we tried to bowl them out.'[189]

Stephen Potter (1900–1969), a popular British author of the mid-twentieth century, produced such humorous self-help classics as *The Theory and Practice of Gamesmanship: Or the Art of Winning Games Without Actually Cheating* (1947) and *One-Upmanship* (1952) where, if you are not one-up you are one-down, a position which identified Potter with one-upmanship, thereby propelling the suffix '-manship' into UK English vocabulary.[190] Written in an under-stated, dead-pan, satirical British style of 1950s humour, it is all about how to use simple psychological game strategies to wrong-foot one's opponent and thereby to become 'one-up' on him or her.[191]

Although by citing Potter's one-upmanship strategy, it may seem as if he was treating a serious subject with inappropriate levity, in fact, Walls was demonstrating his understanding of the need to adopt a suitable psychological approach towards appearing in the witness box, especially where one could not solely rest upon personal status and authority as in the earlier part of the century. His approach signified an appropriate sensitivity to the mid-century British *zeitgeist*, in the country where golf, cricket and soccer may or may not have been invented, but where the first rules probably were, and, through his version of one-upmanship for expert witnesses, he invented 'witnessboxmanship'.[192]

'Good witnessboxmanship consists in knowing how to be one-up without antagonising the court in becoming so.'[193] Clearly, although he would never have cited as light-hearted a source as Potter, Spilsbury was a master of one-upmanship, no less witnessboxmanship. Importantly, Walls understood that witnessboxmanship was *performative*. He suggested two variants, and his descriptions of these make it clear that he was a master of the game and knew exactly how to act. Either the witness had to be so well informed about his subject and able to answer every question in such a self-assured way that counsel 'is willy-nilly forced into the position of one humbly seeking information from the master'; Spilsbury could reasonably be credited with adopting this approach.[194] The second alternative was to adopt a humbler persona of 'honest, struggling craftsman', making it look as though one is aware of one's lack of knowledge, thereby taking a more subordinate role in the common attempt to seek justice.[195]

However, 'witnessboxmanship' was a far from simple game, and Walls was well aware that his strategy could back-fire. The problem with the first approach was that counsel may be better than the witness at this game and it might not be possible to get 'one-up' on him. In any case, adopting this style of witnessing may make the witness appear arrogant which would irritate jury and judge. The second approach could make the witness appear more on the court's wavelength, but if he does not project expert authority, he may not appear sufficiently knowledgeable and could come over as 'bumbling'. Walls did not recommend an approach in the middle; rather, he suggested switching from one approach to the other, picking up a sense of what is required and adapting one's performance as the cross-examination proceeded.[196] It was a skilled dramatic performance.

Walls also suggested that giving the cross-examining counsel more than he asks for could be a useful approach. For instance, suppose there were some discrepancies between the photograph of a supposedly matching tool and a mark where an exact fit has been alleged. The expert witness knew, of course, that this was due to the difficulty of representing three-dimensional details in two dimensions, and he also knew that the lawyer hoped to draw the witness into an overly technical explanation, thereby confusing the jury and making the expert appear less definite about the fit than he initially appeared to be. Spotting this possibility, the witness could then claim that he might have been incompetent in taking the photo, leaving counsel the choice of either accusing the witness of not taking the photo competently, which the court would not accept, as he had already demonstrated his expertise, or of rejecting this suggestion, which would then put the witness one-up and counsel one-down for trying to make a fuss over nothing.[197] The point was not to give counsel a chance to tell the witness he was wrong; rather, the witness had to get the upper hand by letting the court know that he accepted he might be wrong. It was all very clever stuff, and clearly Walls was a master of the art of 'witnessboxmanship'.

Performance in the witness box was, then, an important part of the forensic scientist's toolbox, as until a change in legislation in 1967, forensic scientists made frequent appearances in court to defend their expert evidence.[198] However, this was to alter with the new legislation and increasing case loads. For example, the Metropolitan Police Laboratory dealt with 696, 2988 and 22,783 cases in 1950, 1960 and 1970, respectively.[199] Apart from any other reason, the steep increase in cases would have made it impossible to have significant court representation of expert witnesses; written evidence had to prevail when it was permitted by the new Criminal Justice Act (1967).

However, for all the apparently negative aspects of expert witnesses' appearances in the witness box, important opportunities to present and explain scientific evidence were lost, something that became evident when expert witnesses' appearances in the witness box were considerably less frequent. This was a matter of concern to Walls's successor to the post of Director of the Metropolitan Laboratory, Raymond Williams, appointed in 1968. Although Williams was speaking in 1983, which was somewhat after the focus of the present work, nevertheless he was reflecting on the earlier system, where the expert witness was much more likely to be present in court, in contrast to the later system which relied heavily on written evidence. A comparison usefully sheds light on the advantages of the earlier system. Williams argued that forensic evidence, which took time and effort to prepare, could actually be wasted.[200] The fault could, of course, lie with the investigating officer who might not understand the significance of the forensic evidence. This could be compounded when the case appeared in court. Even if the evidence was compelling in the forensic scientist's opinion, the prosecution's counsel might fail to understand its significance or might wish to pursue another line of cross-examination. In Williams's opinion, the forensic laboratory did not have a close relationship with the barrister. A pre-trial conference could be requested, but there might not be time. The barrister could be pressed for time. '[I] get the impression that the chap picks up his papers two days before the trial and walks into court with a written statement from the Senior Investigating Officer'.[201]

Williams estimated that with the introduction of the Criminal Justice Act of 1967, which made written statements from witnesses admissible in England and Wales, the proportion of cases where an expert witness appeared in court plummeted to less than a tenth of what it was before. It was extremely difficult to express nuances on a written page. If evidence was verbal, the expert witness had the opportunity to amplify it and 'change the whole tenor of the message being transmitted'.[202] The value of verbal evidence was brought home by the example of a visit of Lord Justice Lane to the Metropolitan Laboratory the day after one of its scientists had been presenting hair evidence in court.[203] Williams deliberately put Lord Lane with the scientist, who showed him the relevant slides. At the end of the visit, Lane remarked on how useful this was to give 'a far clearer idea of the case than the bits of paper put in front of him'.[204] Defence counsel knew that if they just accepted the witness statement, it would have much less impact than a skilled expert witness presenting evidence in court, and this reinforced the importance of a good written statement which would not damage the prosecution's case. However, as Williams noted, with an ever-growing case load, there was no going back to the system of appearing in court in every case; the system would grind to a halt within a month of the beginning of the fiscal year. Not surprisingly, Williams regarded the adversarial English system as 'not a good bedfellow with expert evidence'.[205]

Conclusion

One strategy for completing a complicated jigsaw is to find the edges of the picture, to complete the border of the jigsaw before filling in the middle pieces. This chapter attempts to piece together the jigsaw border, a border that runs from at least the seventeenth century to the middle of the twentieth, with the development of the relationship of science and law, facts, probabilistic reasoning, expert witnessing and the performances of scientists in the witness box, the problems of the adversarial system and strategies to cope with these problems and the development of the medical and scientific professions. However, there are many jigsaw pieces to be placed between the edges, and this is the job of succeeding chapters. So far, other than the descriptions which attach to particular legal cases, I have said relatively little about crime and criminals. The focus has been on how the expert witness handled appearances in the law court. Given that forensic science is largely concerned with scientific approaches for the support of criminal justice, we must now step back to consider ways in which criminal and criminality were identified, how a crime scene could be managed and how this was influenced by the apparatus of scientific policing and detection and the role of forensic science in literature, particularly detective literature; all these are elements in the development of forensic science in Britain. The next chapter winds the clock back into the nineteenth century to consider the development of scientific criminology and criminalistics, and how these were utilized in the characterization and identification of the criminal and how these set the scene for the development of forensic science in Britain.

Notes

1 B. Shapiro, *Probability and Certainty in Seventeenth-Century England: A Study of the Relationships Between Natural Science, Religion, History, Law and Literature*, Princeton, NJ: Princeton University Press, 1983.
2 S. Shapin, *A Social History of Truth: Civility and Science in Seventeenth Century England*, Chicago, IL: University of Chicago Press, 1994; S. Shapin and S. Schaffer, *Leviathan and the Air Pump: Hobbes, Boyle and the Experimental Life*, Princeton, NJ: Princeton University Press, 1985.

3 Shapiro, *Probability and Certainty.*
4 C. Dickens, *Hard Times*, Harmondsworth: Penguin, 1969 (originally published 1854), p. 47.
5 B. Shapiro, *A Culture of Fact: England 1550–1720*, Ithaca, NY and London: Cornell University Press, 2000.
6 Ibid., p. 2.
7 M. Hunter, *The Royal Society and Its Fellows, 1660–1700: The Morphology of an Early Scientific Institution*, Chalfont St Giles, Bucks: British Society for the History of Science, 1982.
8 Shapiro, *A Culture of Fact*, p. 4.
9 Ibid., p. 9.
10 Ibid., p. 30.
11 Ibid., p. 320; Shapin and Schaffer, *Leviathan and the Air Pump.*
12 Shapiro, *A Culture of Fact*, p. 5.
13 Ibid., p. 112.
14 Ibid., pp. 139–140.
15 C.A.G. Jones, *Expert Witnesses: Science, Medicine, and the Practice of Law*, Oxford: Clarendon Press, 1994, p. 6.
16 Ibid., p. 10.
17 R. Smith, 'Forensic pathology, scientific expertise, and the criminal law', in R. Smith and B. Wynne (eds), *Expert Evidence: Interpreting Science in the Law*, London and New York: Routledge, 1989, 56–92, p. 64; K. D. Watson, *Forensic Medicine in Western Society: A History*, Abingdon: Routledge, 2011, p. 49.
18 Smith, 'Forensic pathology', p. 65.
19 Ibid., p. 66.
20 Ibid., p. 67.
21 Shapiro, *Probability and Certainty.*
22 Ibid. p. 3.
23 Ibid.
24 S. Shapin, *The Scientific Life: A Moral History of a Late Modern Vocation*, Chicago, IL and London: University of Chicago Press, 2008, p. 3.
25 Shapiro, *Probability and Certainty*, pp. 9–10.
26 S. Shapin, *The Scientific Revolution*, Chicago, IL and London: University of Chicago Press, 1996, p. 102, emphasis in the original.
27 Smith, 'Forensic pathology', p. 70.
28 Ibid., p. 71.
29 Shapin and Schaffer, *Leviathan and the Air Pump*; Shapiro, *Probability and Certainty;* Shapiro, *A Culture of Fact.*
30 Shapin and Schaffer, *Leviathan and the Air Pump*; Hunter, *The Royal Society*, p. 1.
31 Shapin and Schaffer, *Leviathan and the Air Pump*, p. 25.
32 Ibid., p. 60.
33 For replication of scientific experiments, see H. M. Collins, *Changing Order: Replication and Induction in Scientific Practice*, London: Sage, 1983. For administrative objectivity, see M. Lynch, S. A. Cole, R. McNally and K. Jordan, *Truth Machine: The Contentious History of DNA Fingerprinting*, Chicago, IL and London: University of Chicago Press, 2008.
34 For forensic medicine, see, e.g., A. S. Taylor, *The Principles and Practice of Medical Jurisprudence*, London: Churchill, 1865; for forensic science, see, e.g., A. Lucas, *Forensic Chemistry*, London: Edward Arnold, 1921. The first three English-language editions of Gross's textbook were, respectively: J. Adam and J. C. Adam, *Criminal*

Investigation A Practical Handbook for Magistrates, Police Officers and Lawyers, Madras: A. Krishnamachari, 1906; J. Adam and J. C. Adam, *Criminal Investigation A Practical Handbook for Magistrates, Police Officers and Lawyers*, London: The Specialist Press, 1907 (reprint of 1906 edition); and J. C. Adam, *Criminal Investigation A Practical Textbook for Magistrates, Police Officers and Lawyers*, London: Sweet and Maxwell, 1924.

35 K. Simpson, *Taylor's Principles and Practice of Medical Jurisprudence*, London: J. & A. Churchill, 1965, 2 vols., 12th edition.

36 See the example relating to John Glaister, Jnr later in this chapter.

37 S. Shapin, *A Social History of Truth: Civility and Science in Seventeenth Century England*, Chicago, IL: University of Chicago Press, 1994.

38 Shapiro, *A Culture of Fact*, p. 118.

39 Watson, *Forensic Medicine*, p. 47.

40 Jones, *Expert Witnesses*, p. 25.

41 D. M. Dwyer, 'Expert evidence in the English civil courts, 1550–1800', *The Journal of Legal History*, 2007, 28 (1): 93–118.

42 Ibid., p. 118.

43 Ibid.

44 Jones, *Expert Witnesses*, pp. 21–22.

45 T. Golan, 'The history of scientific expert testimony in the English courtroom', *Science in Context*, 1999, 12 (1): 7–32, p. 11.

46 Ibid., p 12.

47 S. Arapostathis, and G. Gooday, *Patently Contestable: Electrical Technologies and Inventor Identities on Trial in Britain*, Cambridge, MA and London: MIT Press, 2013, p. 25; Jones, *Expert Witnesses*, p. 23; Watson, *Forensic Medicine*, p. 47.

48 J. L. Mnookin, 'Idealizing science and demonizing experts: An intellectual history of expert evidence', *Villanova Law Review*, 2007, 52 (101): 763–801, p. 770.

49 V. Berridge, 'Health and medicine', in F.M.L. Thompson (ed), *The Cambridge Social History of Britain 1750–1950: Volume 3: Social Agencies and Institutions*, Cambridge: Cambridge University Press, 1990, 171–242, p. 173.

50 Watson, *Forensic Medicine*, p. 4.

51 P. Corfield, *Power and the Professions in Britain 1700–1850*, London and New York: Routledge, 1995, p. 153.

52 Berridge, 'Health and medicine', p. 176; Corfield, *Power and the Professions*, p. 146.

53 Corfield, *Power and the Professions*, p. 149.

54 Berridge, 'Health and medicine', p. 177; Corfield, *Power and the Professions*, p. 151.

55 Corfield, *Power and the Professions*, p. 137.

56 A. J. Youngson, *The Scientific Revolution in Victorian Medicine*, London: Croom Helm, 1979.

57 A. Adam, *Spontaneous Generation in the 1870s: Victorian Scientific Naturalism and Its Relationship to Medicine*, unpublished PhD thesis, CNAA, Sheffield City Polytechnic, 1989.

58 Ibid.

59 M. J. Peterson, *The Medical Profession in Mid-Victorian London*, Berkeley and Los Angeles: University of California Press, 1978.

60 C. Lawrence, 'Incommunicable knowledge: Science, technology and the clinical art in Britain 1850–1914', *Journal of Contemporary History*, 1985, 20 (4): 503–520.

61 C. A. Russell, N G. Cole and G. K. Roberts, *Chemists by Profession: The Origins and Rise of the Royal Institute of Chemistry*, Milton Keynes: Open University Press/Royal Institute of Chemistry, 1977, pp. 88, 128.

62 Ibid., p. 128.
63 Ibid., p. 94.
64 R. Bud, and G. K. Roberts, *Science Versus Practice: Chemistry in Victorian Britain*, Manchester and Dover, NH: Manchester University Press, 1984, p. 87.
65 F. M. Turner, *Between Science and Religion: The Reaction to Scientific Naturalism in Late Victorian England*, New Haven, CT: Yale University Press, 1974, p. 8.
66 H. Perkin, *The Rise of Professional Society: England Since 1880*, London and New York: Routledge, 1989, p. xii.
67 Bud and Roberts, *Science Versus Practice*, p. 96.
68 Ibid.
69 B. Dyer, and C. A. Mitchell, *The Society of Public Analysts and Other Analytical Chemists: Some Reminiscences of Its First Fifty Years and Review of Its Activities*, Cambridge: Heffer, 1932.
70 Chapter 5 charts the relationship of public analysts to forensic work in more detail.
71 D.S.L. Cardwell, *The Organisation of Science in England: A Retrospect*, Melbourne, London, Toronto: Wm Heinemann, 1957.
72 Gender was even more significant than class. There was very little opportunity for a scientifically inclined woman from the middle classes to make a living from science and even less opportunity for a working-class woman to achieve a career as a professional scientist. See S. K. Muka, 'Portrait of an outsider: Class, gender, and the scientific career of Ida M. Mellen', *Journal of the History of Biology*, 2013, available at http://link.springer.com/article/10.1007/s10739–013–9354-z, accessed 20 April 2013; M. W. Rossiter, *Women Scientists in America: Struggles and Strategies to 1940, Vol. 1*, Baltimore, MD: The Johns Hopkins University Press, 1982. This is reflected in the history of forensic sciences which, until somewhat after the middle of the twentieth century, was largely a world without women. A notable exception is Margaret Pereira, who joined the Metropolitan Police Laboratory in 1947 and was Controller of the Forensic Science Service from 1982–1988. See "Pereira, Margaret', *Who's Who 2014*, A & C Black, an imprint of Bloomsbury Publishing plc, 2014, Oxford University Press, 2014; online edn, Nov 2014, available at http://www.ukwhoswho.com/view/article/oupww/whoswho/U30563, accessed 2 May 2015..
73 Cardwell, *The Organisation of Science*.
74 E. Schatzberg, 'From art to applied science', *Isis*, 2012, 103 (3): 555–563.
75 Some incumbents of university chairs held their posts for decades; e.g., William Thompson, Baron Kelvin of Largs, held his position as Professor of Natural Philosophy at the University of Glasgow for over fifty years.
76 G. Gooday, '"Vague and artificial": the historically elusive distinction between pure and applied science', *Isis*, 2012, 103 (3): 546–554, p. 551.
77 S. Clarke, 'Pure science with a practical aim: The meanings of fundamental research in Britain, circa 1916–1950', *Isis*, 2010, 101 (2): 285–311, p. 289.
78 E.g., see W. M. Else and J. M. Garrow, *The Detection of Crime: An Introduction to Some Methods of Scientific Aid in Criminal Investigation*, London: The Police Journal, 1934.
79 M. J. Saks, 'Banishing ipse dixit: The impact of Kumho Tire on forensic identification science', *Washington and Lee Law Review*, 2000, 57 (3): 879–900, available at http://scholarlycommons.law.wlu.edu/wlulr/vol57/iss3/7, accessed 6 January 2015.
80 HMSO, *Home Office, Scientific Aids to Criminal Investigation, Forensic Science Circulars*, No. 1, HMSO, London, 1936.
81 See D. S. Caudill, 'Arsenic and old chemistry: Images of mad alchemists, experts attacking experts, and the crisis in forensic science', *Villanova University School of Law Working Paper Series, Paper 136, 2009*, available at http://digitalcommons.

law.villanova.edu/cgi/viewcontent.cgi?article=1140&context=wps, accessed 3 January 2015; J. L. Mnookin, 'Scripting expertise: The history of handwriting identification evidence and the judicial construction of reliability', *Virginia Law Review*, 2001, 87 (8): 1723–1845; National Research Council, National Academy of Sciences (NRC-NAS), *Strengthening Forensic Science in the United States: A Path Forward*, Washington DC: National Academies Press, 2009.

82 Caudill 'Arsenic and old chemistry', p. 32. Caudill may be referring to Kuhn's characterization of 'normal science'. See T. S. Kuhn, *The Structure of Scientific Revolutions*, Chicago, IL: University of Chicago Press, 1962.

83 Saks, 'Banishing ipse dixit', p. 882.

84 NRCNAS, *Strengthening Forensic Science*.

85 Saks, 'Banishing ipse dixit'.

86 R. Smith, *Technical/Scientific, Social and Industrial Expertise and Legal Frameworks 1983–84 (SERC/ESRC)*, Recorded interviews with leading practitioners and commentators on legal medicine and forensic science conducted by Dr Roger Smith, Dept. of History, University of Lancaster, Raymond Williams, 21 December, 1983, Wellcome Library, London, 1983–1984.

87 Mnookin, 'Scripting expertise', p. 1727.

88 Ibid., p. 1732.

89 Ibid.

90 Ibid., p. 1742.

91 Ibid., p. 1818.

92 Golan, 'The history', p. 15; C. Hamlin, 'Scientific method and expert witnessing: Victorian perspectives on a modern problem', *Social Studies of Science*, 1986, 16 (3): 485–513, pp. 488–489.

93 Arapostathis and Gooday, *Patently Contestable*, p. 25.

94 Golan, 'The history', pp. 15–16.

95 G. Gooday, 'Liars, experts and authorities', *History of Science*, 2008, 46 (4): 431–456; Hamlin, 'Scientific method; Mnookin, 'Idealizing science'.

96 Hamlin, 'Scientific method', p. 486; Smith, 'Forensic pathology'.

97 Hamlin, 'Scientific method', p. 488.

98 Ibid., p. 491.

99 Ibid., p. 502.

100 Ibid., p. 497.

101 Golan, 'The history', p. 26.

102 Mnookin, 'Idealizing science', p. 771.

103 Ibid., p. 782.

104 Gooday, 'Liars', p. 446.

105 T. Golan, *Laws of Men and Laws of Nature: The History of Scientific Expert Testimony in England and America*, Cambridge, MA and London: Harvard University Press, 2004, p. 110.

106 I. Burney, *Poison, Detection, and the Victorian Imagination*, Manchester: Manchester University Press, 2006, p. 116.

107 Burney, *Poison*; Golan, *Laws of Men*, pp. 97–100; K. D. Watson, *Poisoned Lives: English Poisoners and Their Victims*, London and New York: Hambledon and London, 2004, pp. 103–104.

108 Golan, *Laws of Men*, p. 97. Sir James Stephen was a leading Victorian jurist and legal practitioner. See T. Ward, 'A mania for suspicion: Poisoning, science and the law', in J. Rowbotham and K. Stevenson (eds), *Criminal Conversations: Victorian Crime, Social Panic, and Moral Outrage*, Columbus: Ohio State University Press, 2005,140–156.

109 Simpson, *Taylor's Principles*.
110 Golan, *Laws of Men*, p. 99.
111 Ibid. p. 100.
112 See Golan, *Laws of Men* and, especially Burney, *Poison*, for a full description of the Palmer trial.
113 Watson, *Forensic* Medicine, p. 67.
114 Golan, *Laws of Men*; Mnookin, 'Idealizing science'.
115 For Frye and Daubert tests, see T. Golan, 'Revisiting the history of scientific expert testimony', *Brooklyn Law Review* (2008) 73 (3): 879–942; S. Jasanoff, *Science at the Bar: Law, Science and Technology in America*, Cambridge, MA and London: Harvard University Press, 1995, pp. 61–64.
116 J. Fisher, *Forensics Under Fire: Are Bad Science and Dueling Experts Corrupting Criminal Justice?* New Brunswick, NJ: Rutgers University Press, 2008; P. Huber, *Galileo's Revenge: Junk Science in the Courtroom*, New York: Basic, 1991.
117 Smith, 'Forensic pathology', p. 72.
118 Ibid., p. 248. Forensic confirmation bias is now recognized as a significant problem in relation to forensic evidence; e.g., see S. D. Charman, 'The forensic confirmation bias: A problem of evidence integration, not just evidence evaluation', *Journal of Applied Research in Memory and Cognition*, 2013, 2 (1): 56–58.
119 J. Grant, *Science for the Prosecution*, London: Chapman and Hall, 1941.
120 C. A. Mitchell, 'Review of *Science for the Prosecution*', *The Analyst*, 1942, 67: 154.
121 C. A. Mitchell, *Science and the Criminal*, London: Isaac Pitman & Sons, 1911; C. A. Mitchell. *The Scientific Detective and the Expert Witness*, Cambridge: W. Heffer & Sons Ltd, 1931; C. A. Mitchell, *A Scientist in the Criminal Courts*, London: Chapman & Hall Ltd, 1945. Mitchell *The Scientific Detective* was the reissue of a 1923 edition.
122 Mitchell, *Science and the Criminal*, p. 13.
123 Ibid.
124 Ibid., p. 14. Note that this book was published in 1911.
125 Ibid.
126 Ibid.
127 B. Prainsack and M. Kitzberger, 'DNA behind bars: Other ways of knowing forensic DNA technologies', *Social Studies of Science*, 2009, 39 (1): 51–79.
128 J. McCafferty, *Mac, I've Got a Murder*, London: Arthur Barker Ltd., 1975, p. 136.
129 Mitchell, *A Scientist*, p. 33.
130 Ibid., p. 34.
131 Smith, 'Forensic pathology'.
132 S. Smith, *Mostly Murder*, London: Harrap, pp. 152–153.
133 Ibid., p. 153.
134 For Spilsbury's influence, see I. Burney, and N. Pemberton, 'Bruised witness: Bernard Spilsbury and the performance of early twentieth-century English forensic pathology', *Medical History*, 2011, 55: 41–46, p. 41. See K. D. Watson, *Dr Crippen*, Kew, Richmond: The National Archives, 2007 for a detailed analysis of materials relating to the Crippen case held in the National Archives.
135 Watson, *Dr Crippen*, p. 50.
136 A. Rose, *Lethal Witness: Sir Bernard Spilsbury Honorary Pathologist*, Chalfont, Stroud: Sutton, 2007, pp. xix–xx, p. 164.
137 E.g., see K. Simpson, *Forty Years of Murder: An Autobiography*, London: Harrap, 1978, p. 26.

138 Mnookin, 'Scripting expertise'.
139 Rose, *Lethal Witness*, p. 5.
140 Burney, and Pemberton, 'Bruised witness' p. 57.
141 Rose, *Lethal Witness*, p. xix.
142 M. A. Green, 'Is Sir Bernard Spilsbury dead?' in A. R. Brownlie (ed.), *Criminal Investigation Art or Science?*, Edinburgh and London: Scottish Academic Press, 1984, 23–26, p. 25.
143 Ibid.
144 Ibid., p. 26.
145 Rose, *Lethal Witness*, p. 41.
146 M. Hastings, *The Other Mr Churchill: A Lifetime of Shooting and Murder*, London: George Harrap & Co. Ltd, 1963, p. 131.
147 T. A. Warlow, *Firearms, the Law and Forensic Ballistics*, Boca Raton, FL: CRC Press, 2005, p. 51.
148 Hastings, *The Other Mr Churchill.*
149 Ibid., frontispiece.
150 Rose, *Lethal Witness*, p. 42.
151 Smith *Mostly Murder*, p. 143.
152 Ibid., p. 145. At the time suicide was a crime in England and Wales although not in Scotland – nevertheless considerable stigma was attached to suicide.
153 Ibid. p. 143.
154 Ibid., p. 146.
155 Hastings, *The Other Mr Churchill*, p. 108.
156 Smith, *Mostly Murder*, p. 144.
157 Ibid.
158 Rose, *Lethal Witness*, pp. 147–148.
159 Hastings, *The Other Mr Churchill*, p. 110.
160 Rose, *Lethal Witness*, p. 148.
161 Ibid.; also see Hastings, *The Other Mr Churchill*, p. 124.
162 For an explanation of 'not proven' in Scottish law, see P. Duff, 'The Scottish criminal jury: A very peculiar institution', *Law and Contemporary Problems*, 1999, 62 (2): 173–201, p. 193: 'the verdict of "not guilty" is thought to mean that the accused definitely did not commit the crime, that is, it is a positive declaration of innocence, whereas the verdict of "not proven" is thought to imply solely that the accused's guilt has not been conclusively demonstrated.': Rose, *Lethal Witness*, p. 151.
163 Smith, *Mostly Murder*, p. 144
164 Ibid.
165 Ibid., p. 162.
166 Ibid.
167 J. Glaister Jnr., *Final Diagnosis*, London: Hutchinson, 1964, p. 52.
168 Ibid. For precipitin test, see J. Glaister Jnr., 'Some results of recent medico-legal research in the examination of blood-stains and hairs', *The Police Journal*, 1928, 1 (1): 62–77.
169 Glaister Jnr., *Final Diagnosis*, p. 52.
170 J. B. Firth, *A Scientist Turns to Crime*, London: William Kimber, 1960, p. 19.
171 H. J. Walls, *Expert Witness: My Thirty Years in Forensic Science*, London: John Long, 1972.
172 Ibid., p. 176.
173 Ibid.

174 Ibid., p. 177.
175 Ibid.
176 F. G. Tryhorn, 'Scientific aids in criminal investigation Part I', *The Police Journal*, 1936, 9 (1): 33–41, p. 34.
177 Walls, *Expert Witness*, p. 180.
178 Smith, *Technical/Scientific*, Interview with Hamish J. Walls, 2 June 1983.
179 Ibid. Interview with Raymond Williams, 21 December, 1983.
180 Walls, *Expert Witness*, p. 181.
181 Ibid., p. 183.
182 Ibid., p. 185.
183 Ibid. Walls weighed up the possibility of assigning numerical values to decisions about guilt or innocence, even the possibility of 'Mr Justice Computer' (see Ibid). Only a decade later, part of the UK's academic community was in the throes of trying to formalize legal decision making by means of a computer in 'expert systems' as part of the UK's fifth-generation computing project – see the Alvey/DHSS demonstrator project. (See T.J.M. Bench-Capon (ed.), *Knowledge-Based Systems and Legal Applications*, London: Academic Press, 1991.) For the Sage expert system for forensic science developed in relation to the Forensic Science Service, see C. Joyce, 'The clues to forensic science found by computer', *New Scientist*, 27 September, 1984, p. 8.
184 Walls, *Expert Witness*, p. 185.
185 Ibid., p. 186.
186 Ibid., p. 188.
187 Ibid.
188 Smith, 'Forensic pathology', p. 72.
189 Smith, *Technical/Scientific*, Interview with Hamish J. Walls, 2 June 1983.
190 S. Potter, *The Theory and Practice of Gamesmanship: Or the Art of Winning Games Without Actually Cheating*, London: Rupert Hart-Davis, 1947; S. Potter, *One-Upmanship. . . Being Some Account of the Activities and Teaching of the Lifemanship Correspondence College of One-Upness and Gameslifemastery*, London: Rupert Hart-Davis, 1952.
191 B. Lowrey, 'The timelessness of Stephen Potter's gamesmanship', *The Virginia Quarterly Review: A National Journal of Literature and Discussion*, 1993, 69 (4): 718–726.
192 Walls, *Expert Witness*, p. 189.
193 Ibid.
194 Ibid.
195 Ibid.
196 Ibid., p. 190.
197 Ibid.
198 For Criminal Justice Act (1967) see http://www.legislation.gov.uk/ukpga/1967/80, accessed 3 May 2015.
199 R. F. Coleman, and H. J. Walls, 'The evaluation of scientific evidence', *The Criminal Law Review*, 1974: 276–288: p. 276.
200 Smith, *Technical/Scientific*, Interview with Raymond Williams, 21 December, 1983.
201 Ibid.
202 Ibid.
203 Lord Justice Lane was Lord Chief Justice of England from 1980 to 1992.
204 Smith, *Technical/Scientific*, Interview with Raymond Williams, 21 December, 1983.
205 Ibid.

2 The influence of scientific criminology and criminalistics

Introduction

Chapter 1 set the scene by considering the shaping of epistemological terms in science and in law, the emergence of the expert scientific witness and the tactics employed by twentieth-century scientific witnesses to present their evidence effectively under cross-examination in the courtroom. However, before the court case comes the identification of a crime, and the suspect must be linked to the crime.[1] This and following chapters step out of the courtroom to consider how science and scientific ideas were employed in the fight against crime. Strategies to identify the criminal and to connect the criminal to the evidence influenced the growth of the new forensic science that was developing in the early years of the twentieth century in Britain.

Scientific criminology and criminalistics were the main candidates for a 'scientific' approach towards crime at the turn of the nineteenth century. In the second half of the nineteenth century in the UK, although forensic medicine was a recognized discipline, forensic science was not.[2] It is important not to make too fine a distinction in this period, as medico-legal experts undertook a wide range of scientific analyses well into the twentieth century and, as we saw in the previous chapter, some were even performing ballistics experiments and presenting the results in court. Toxicological analyses for poisoning cases could arguably be considered the first type of forensic science. There were also many legal cases involving specialist scientific and engineering expertise for insurance and manufacturing purposes in fraud and counterfeiting. Apart from toxicology, a specific body of scientific knowledge and techniques for analysing crime scenes and trace evidence, whether new or borrowed from existing science, was starting to coalesce.[3] Although it was by no means the only factor, scientific criminology can be seen as one of the major influences on the development of scientific perspectives on crime and criminals.[4]

This chapter considers the influence of scientific criminology, from the UK and beyond, in creating some of the conditions which were to support the growth of forensic science. Scientific criminology focused on the criminal rather than the crime; indeed, a recurring criticism is that it appeared to say little about how a criminal is connected to a specific crime – ultimately, this was a weakness.

Nevertheless, the idea that criminality could be observed and measured in an objective way was extremely powerful. Statistical analyses of *average man* formed the backdrop for attempts to uncover the markers of individual criminality, in particular, physical signs of criminality.[5] The job of scientific criminology involved identifying the habitual criminal, recidivist or 'born criminal', given the assumption that there were criminal classes or even criminal races. So it was a question of identifying whether someone had criminal propensities, identified mainly though physical indicators, an evolutionary and hereditary approach which said little about the social conditions where criminal behaviour was fostered. Scientific criminology looked to the category rather than the individual.

The *criminalistic* approach of Hans Gross (1847–1915), presenting something of a contrast to scientific criminology, has only recently received the attention it deserves from criminological and historical scholars.[6] Although Gross distinguished his system from that of scientific criminology, he, too, was much exercised about criminal psychology alongside his focus on the crime scene, how to manage it and how to apply scientific analyses to the evidence gathered. However, he was not a scientist, having been trained in law.[7] It is tempting to view Gross as far ahead of his time because of his emphasis on the crime scene; this is the part of his legacy that tends to be remembered by those who enthusiastically hail his pioneering role. However, he was also *of* his time, especially in his demonizing of nomadic people. The latter can, of course, be forgotten, as indeed it was in later English-language editions of his major work, *Handbuch für Untersuchungsrichter* (Handbook for Examining Magistrates).[8] However, considering the whole ambit of Gross's work shows his hard line on what he considered to be criminal groups was more in tune with scientific criminology than might first appear. His views on the criminality of wandering tribes were based on an evolutionary view that tended to equate criminality with those who were deemed 'savage'. This is not so distant from Lombroso, one of the main apologists for scientific criminology. Gross's methods can be seen as something of a bridge between scientific criminology and the new forensic sciences that were developed in the twentieth century to analyse the material collected from crime scenes. Gross was undoubtedly significant.[9] The appeal of his work lay in his detailed consideration of the psychology of criminals and witnesses and, particularly, in his system of managing and preserving evidence from the crime scene to identify the perpetrator. Gross told you what to do at the crime scene. Lombroso, Galton and others involved in scientific criminology did not. His work, at least on the management of the crime scene, was regarded as being so far in advance of its time that it is still referenced by modern criminalists and forensic scientists, who name him the 'father of forensic investigation'.[10] He has achieved mythical status.

The genealogy of English translations of his influential *Handbuch* signals that it had a significant trajectory in policing the empire. The first English translations of this work were undertaken by senior British lawyers in Bengal, importing local examples and remaking the book into a kind of handbook of colonial scientific policing which then spread to the UK and USA.[11] Clearly, there are some things about Hans Gross that posterity would prefer to forget. Gross had plenty to say about the criminal character, the nature of *Gauner*, or criminals and their habits and tricks, as well as the characters of witnesses.[12] Successive English editions of his *Handbuch* were to achieve an appropriate level of forgetfulness of his more problematic opinions as they were gradually edited out over the fifty-five years that his book remained in print, as we shall see below.

The meaning of 'scientific'

When we talk of scientific criminology, scientific policing or, indeed, scientific anything, we must consider what 'scientific' means. The epithet should not be taken for granted. Rather than attaching to a particular scientific technique or disciplinary area, it was often used as a rhetorical device to signal a claim for being modern, objective, organized and, importantly, in control, thereby to be contrasted with older approaches which could then be regarded as 'unscientific' and, therefore, unsatisfactory. Of course, there were plenty of negative connotations attached to the term 'scientific' in the nineteenth century. Apart from the way that early to mid-nineteenth century geology and natural history were considered in some quarters to be mounting an attack on religion, one only has to think of Mary Shelley's *Frankenstein* to see a terrifying vision of science out of control, albeit in fictional form. Science and technology apparently fuelling the relentless engine of industry and capitalism may have been modern and progressive, but they were also exploitative.[13]

'Scientific' acted as a black box term, the meaning of which should be opened up to inspection. Lynch argues that the label scientific in relation to criminology helped to deflect inspection and criticism of the discipline in its early forms.[14] It may have been positive in its claims to modernity and progressive thinking, but it was also negative in its tendency to objectivize and regulate. Scientific criminology inherited the negative side of Enlightenment thinking, where to maintain the human freedoms emphasized by Enlightenment scholarship, conditions of control had to be established.[15] Scientific criminology's focus on the classificatory 'body sciences' which created the human body as a measurable entity 'expanded the tendency to treat humans as mathematical abstractions . . . objects of investigation'.[16] The darker side of the rise of scientific criminology lay in its mechanisms for naming and controlling those who were to be seen as the dangerous classes. Hence, concentrating on scientific criminology's emphasis on scientific explanations should not divert us from understanding it 'as part of the apparatus of the "science of oppression"'. It was one of the ways in which the newly emergent moneyed merchant classes were able to 'legitimize control of the dangerous classes that had spread rapidly throughout Europe during the 17th, 18th and 19th centuries, and in the U.S. in the late 19th and 20th centuries and inculcate discipline'.[17]

Hence, labelling a field 'scientific' was imbued with political meaning, a meaning which could be far from positive and emancipatory. Designating a field 'scientific' involved an emphasis on the perceived objectivity, measurement, exactness, and cultural authority of the sciences and on the moral authority of the scientist as a dispassionate and objective observer of the natural world. Thus, it was the new scientific criminology which gained ascendancy at a time when positivist sociology was becoming popular, based on the view that human behaviour can be studied and observed, and ultimately controlled, according to natural laws. The last decades of the nineteenth century witnessed a substantial growth in a programme of activity in Europe and North America which succeeded in establishing the discipline of scientific criminology in the criminal justice system, including the publication of huge numbers of books and many conferences and congresses, promoting a widespread movement 'which pressed the claims of criminology upon the legislatures and penal institution of virtually every western nation'.[18]

Enlightenment criminology

The new scientific criminology arose in the wake of Enlightenment thinking on how criminals should be treated. The Enlightenment view entailed a move towards more humanitarian thinking about crime and punishment and away from older ideas on sin and torture. The Italian philosopher Cesare Beccaria (1738–1794) and his book *On Crimes and Punishments* (1764) are associated with the Classical school of criminology: 'the first naturalistic explanation of crime – one that did not base itself on concepts of sin and demonology'.[19] Instead, this approach emphasized free will, contract theory, laws made by the agreement of free men, utility, which would find expression in Bentham's 'greatest good for the greatest number' and an emphasis on reason.[20] Such a view rested on the rationality of owning property and on societal inequality. Beccaria was advancing humane views but did not have 'the scientific framework to embrace a truly "rational" view of behaviour'.[21] Beccaria's position was conservative, but its appeal lay in offering a humanitarian position with a rational basis for morality without having to adopt radical materialism. He was arguing a case for reform to be achieved within existing social institutions. 'Beccaria's work stands as a conservative monument, the first great effort to cure crime without curing the society which produced it.'[22] His views remained popular until at least the middle of nineteenth century where they were subsumed by the rise of positivism, determinism and scientific criminology.

Foucault in his oft-quoted *Discipline and Punish* (1977), charted the rise of control over individuals through the development of prisons, police and legal hierarchies – social control was to be achieved through the growth of disciplinary society.[23] Discipline was required to produce docile bodies for the industrial age. Prison was a major part of the disciplinary system, but the system also included the network of institutions, such as schools and factories, which watched and controlled members of society. Hence, Foucault argued: 'disciplinary careers' were created in a system which, in the nineteenth century, produced docility and delinquency by the same means.[24] The humanitarian establishments of the nineteenth century created such disciplinary careers from cradle to the grave – someone orphaned at an early age went from nursery to orphanage, to schools, hospitals and almshouses. The system was controlled by the scientific authority of medicine and criminology, and it produced the delinquents it sought to control. Of course, 'disciplinary careers' also included the careers of those who controlled and disciplined as much as those who were disciplined by those in authority. Indeed, those who enforced discipline were themselves disciplined by the structures of control. So medical officers, police, detectives and ultimately forensic scientists were all part of the structure of disciplinary careers. Foucault's views on discipline and control are not just applicable to the tenets of scientific criminology, they are also relevant to the criminalistics approach which sought to manage the crime scene, the witnesses, criminals and the investigating officers whose job it was to gather and manage the scientific evidence in order to bring a prosecution; all were part of the disciplinary system.

Measurement and statistics – the population and the person

Mathematics and measurement were the definitive badges of a scientific discipline's claim to cultural authority; they, too, are tools of control.[25] Measurement was seen as key to an understanding of human behaviour, in particular, criminal behaviour, embodied in the new scientific criminology. In the burgeoning discipline, measurement took a number of forms. These included attempts to measure the human body accurately with rulers and callipers, as in Alphonse Bertillon's (1853–1914) anthropomorphic system (considered in more detail in Chapter 3), which he developed while serving with the police in Paris.[26] Anthropometric measurements offered the promise of uniquely identifying an individual through a set of body measurements which were believed to be unique to the individual and to remain unchanged throughout adult life. Bertillon's work sat alongside the concern, or perhaps even obsession, with the measurement of skulls and faces in the work of Lombroso and Galton.[27] Alongside the interest in physical measurements of individuals, we see the use of population statistics to measure society in general, and criminals in particular. Goring's *The English Convict* offered good examples of the latter.[28]

The rise of scientific criminology also heralded an extension of the function of the state in the management of crime and a promotion of some professional roles over others, 'prison executives, forensic scientists, psychiatrists and other "penological experts". It widened the fields of medicine . . . and sought to effect a shift of power away from the judiciary and towards a non-legal executive staff'.[29] So the development of criminology was, at least to some extent, concerned with professional division of labour and with challenging legal hegemony. The development of forensic science in the UK was part of the business of moving expertise away from the legal profession into state control with the development of the local government public analysts system in the Victorian era to enforce food adulteration laws and, from the 1930s onwards, with the development of the Home Office forensic science laboratories which maintained strong links with the police. All are part of state apparatus, in the broad service of the law, but with a remit and professional staff which were essentially separate from the judiciary.

Adolphe Quetelet (1796–1874), widely recognized as the founding father of criminal statistics, demonstrated, through statistical means, that criminality could be understood as a social fact.[30] He produced a view of social reality in the form of grids and tables which marshalled and organized the original data into a visual representation, thereby rendering social relations visible, in particular, those of crime and criminality even though these were invisible without the statistics and charts.[31] Quetelet applied the statistical approaches he had learned in the field of astronomy to human behaviour, representing such data pictorially in maps, tables and graphs, thereby clearly demonstrating correlations and implied potential causes. The latter were important. It was not just data that were laid bare; the correlations between data items could be made visible and this permitted causes of behaviour to be determined – 'the promise of pure facts'.[32] One such behaviour which these facts could reveal was criminality and its causes. So social causes could be subject to discoverable scientific laws. We take such visual representations for granted nowadays. Indeed, graphical representations are not just the preserve of scientific papers, they are everywhere in our mass media, and we know how to 'read' them. But we need to remember that data was not always presented thus and types of reading and visualization which we now accept without question had to be constructed and achieved, and the ways of seeing associated with such pictures had to be learned.

Graphical representations, as objective representations of fact, are historically and culturally contingent.[33] Quetelet's innovation centred not only on applying statistics to people's behaviour, but also on displaying it in an innovative, often dramatic, way. The graphical, essentially pictorial, way of displaying data quickly gained ascendancy in nineteenth-century science, rapidly becoming an essential part of the armoury of scientific objectivity.[34] By bringing his scientific, statistical reasoning to bear on social statistics, Quetelet was a pioneer in the creation of the modern statistical approach which was born 'from the recombining of scientific and administrative practices that were initially far apart'.[35] This is an important point because much of the story of the rise of scientific criminology, the criminalistic approach and ultimately the organization of forensic science in the UK involved a combination of science, technologies and administrative structures and procedures. Through summing individual, apparently random behaviour, Quetelet demonstrated the statistical regularity of the *average man*.[36] This was based on a normal or Gaussian probability distribution – the typical bell-shaped curve of population data which clusters round a mean value – and on sets of what he called 'moral statistics', in other words data about human populations. Individuality was subsumed under the process of averaging. Coupled with the presumption of scientific laws governing social processes, this militated against Enlightenment notions of individual free will and the idea of the individual entering freely into a social contract.

Quetelet's art can be understood in his connection of probability, statistical observations *and* administrative procedures, with the administrative side running alongside 'pure' science as very much part of the art. Of course, the importance of administration would be seen to good effect in the criminalistic approach of Gross and the development of forensic science in the twentieth century, particularly in terms of the protocols for handling crime scenes, packing material and so on – what Lynch et al. term, 'administrative objectivity'.[37] In Quetelet's moral statistics: 'We are offered a new governing rationality, not a social contract but of a table of weights and measures, of costs and benefits'.[38] Administrative objectivity combines effectively with utilitarianism to average out individual difference.

Importantly, Quetelet moved probabilistic thinking from the earlier legal view of probability where the human dimension was much more explicit, as Chapter 1 describes, towards a scientific, objective view of probability where statistical laws could be derived; in the process, his 'moral statistics' acquired the mantle of objectivity. The older, legal view of probability emphasized human assessment of the likelihood of something happening – the idea that a reasonable person would regard a particular legal 'fact' as likely to a certain degree. In the new statistics, the human element of reasonableness is removed, or at least disguised, in the business of nailing down mathematical precision.

Hence, Quetelet's skill lay in taking the large numbers of statistical records which were increasingly produced by bureaucratic institutions and in creating a new visual language with which to understand them.[39] Understanding the human condition was no longer a question of attending to individual rationality; rather, it was to be found in the averaging of the social group – in other words, society rather than the individual. The philosophy of the Enlightenment emphasized individual rationality and choice based on free will. However, the approach championed by Quetelet says nothing about individual rationality; it is a 'macro' approach. This allowed society to be viewed as a whole, apparently from outside, through the eyes of statisticians or scientists who turn their disinterested and dispassionate gaze upon the social world. Society may be seen as measurable and predictable rather than unstable and unpredictable, and it was the scientist who was to make the measurements. One of the effects of such a view was to achieve separation of the observer from that which is observed. This serves to underline an important characteristic of positivist approaches to science and social sciences, namely the construction of the objective external observer who has access to a stable, objective world that can be measured. The observer neither influences nor is influenced by that external world, and the character of the observer is held to be unimportant. This was an important point for scientific witnessing, as it implied that it was possible for a scientific witness to take an objective view. So the fact that the scientific observer had a bias or a gender or a nationality sank out of consideration. In epistemological terms it was a 'view from nowhere'.[40]

Given the unpredictability of the outcomes of the French Revolution and the fears which this raised on the Continent and in the UK, the possibility of measuring, predicting and controlling societies was clearly appealing in the nineteenth century. The objective measurements which Quetelet produced fed such anxieties and signalled a move away from an Enlightenment view of the individual as a rational being towards a nineteenth-century interpretation which replaced this with 'the normal man, the average of a large number of different men all of whom shared in a sum that exceeded them'.[41] If concerns over the potential irrationalities and seeming unpredictability of social revolutions could be quelled, in some part, by the new statistical knowledge, so, too, could anxieties about that other related great societal worry, crime, be alleviated by the new statistics, as criminal behaviour could be objectively understood by statistical laws. If these laws offered the promise of understanding causes of crime, then the possibility of controlling crime also emerged.

Where the new statistics offered the hope of understanding the *average man* through statistical analysis, anthropological approaches promised a more direct understanding of *criminal man*.[42] The key aspect of this was the measurement of many individuals to determine if there were physically measurable aspects of the body that could be used as an identification of criminality. Phrenology, the measurement of skulls, which were held to be the external manifestation of interior brain functions, was initially an important part of the scientific armoury.[43] Related to phrenology, physiognomy, or the view that a person's moral character could be read from physical appearance, particularly the face, was very influential in the development of the new scientific criminology.[44] Physiognomy was an ancient idea, but it found contemporary expression in the new scientific criminology when it was tied to understandings of evolution and to new technologies such as photography, which offered the potential to capture the image of the *criminal type*.[45] One unfortunate aspect of such a characterization was the idea that someone who had criminal tendencies could be identified in advance; hence the 'stigmata' of the criminal physiognomy could mark someone out as a criminal. Anthropometric knowledge would become as important as statistical knowledge in knowing the criminal.

Quetelet's average man transmuted without difficulty into a kind of criminologically *reasonable man* to be measured against the *criminal man* of Lombroso, which is described later. When this was translated in the courtroom in Victorian times, judges were forced into making anthropological assumptions about normal and reasonable behaviour so the behaviour of a *reasonable man* could be taken as the standard.[46] In particular, the norm of the reasonable man could be used to help the jury decide whether a murder charge could be reduced to manslaughter – this was a very important distinction in times when murder carried the death sentence whereas manslaughter did not.[47] Importing the concepts of the new anthropologically informed criminology into the court had the effect of reinforcing norms of reasonableness and deviance, but also paved the way for further types of forensic evidence and forensic expertise, specifically medical and psychiatric, to be used in court. The question of who had the necessary expertise was, of course, crucial. Not everyone was a scientist, but those who were scientists, particularly those who regarded themselves as criminologists, had a special professional authority to study people who were held to be deviant: 'an institutionally sanctioned authority that characterized both their daily routines and their reflections on the nature of crime . . . Criminologists legitimated their authority through the superiority of their skills of observation and analysis over commonsense approaches to the crime problem'.[48]

Measuring, statistical analysis and graphical representation was one part of the development of scientific criminology. Confidence in the ability of an appropriately qualified external observer to access a stable objective reality was the second, and the third element was evolution. Anthropological approaches, embodied in the new scientific criminology, rested upon a belief in evolution; this is Darwinian evolution, albeit in a 'selective' form. Historians of science are good at showing how what we thought were the big paradigm shifts in science never really happened. Just as it now seems that the Scientific Revolution was not what we thought it was, neither, apparently, was the Darwinian Revolution.[49] Bowler argues that the so-called Darwinian Revolution of the second half of the nineteenth century never took place, and Darwin's view of evolution came into its own in biology in the 1920s and 1930s when combined with the new genetics which offered mechanisms for explaining heredity absent in Darwin's original conception of evolution.[50]

However, evolutionary views were widespread from at least the beginning of the nineteenth century, well before Darwin's research appeared on the scene. These non-Darwinian, developmental views of evolution were, in any case, far more useful to the new scientific criminology than a strictly Darwinian approach. The salient point relates to the abandonment of creationism, not whether there was a precise adherence to a belief in natural selection. Once creationism was discarded, the possibility for development and change emerged. Darwinian evolution eschews teleology; it is all a bit random, pointless and driven by the environment. Although it was not the dominant nineteenth-century framework, nevertheless Darwin's theory acted as a catalyst for the widespread acceptance of evolutionary beliefs. What Bowler terms the 'Non-Darwinian Revolution'

> . . . required the rejection of certain key aspects of creationism but it was non-Darwinian because it succeeded in preserving and modernizing the old teleological view of things . . . the story of nineteenth-century evolutionism centers not on Darwinism . . . but on the emergence of what might be called the 'developmental' model of evolution.[51]

Darwin's variational model of evolution pictured evolution as an irregularly haphazardly branching tree, whereas a developmental or progressive model saw evolution as a progression where humankind is the most advanced state. Such a view was consistent with atavism, where some people were seen as belonging to a less advanced state. Indeed, it is much easier to incorporate views that some people were less civilized under the developmental rather than the Darwinian view of evolution because the developmental view regarded Western civilization as the highest peak of evolution. So the new scientific criminology looked to a developmental or progressive view of evolution rather than the random variations apparently offered by Darwinian evolution. The developmental, teleological view of evolution supported views promoted by scientific criminology that some were born criminal. *Criminal man* could be seen as a less evolved or atavistic human type. Similarly, other races could be seen as less developed, raising the colonial question of civilizing other races, with the potential for criminal behaviour, who were held to be less 'advanced' than those of European descent. This thinking came together in the view that some races were naturally criminal and, indeed, this is something which is evident both in Lombroso's work and in the thinking of Hans Gross in relation to gypsies and the transposition of Gross's views onto supposedly criminal tribes in India.[52]

Galton and composite photography

Francis Galton's (1822–1911) relationship to the British criminal justice system is interesting, not least because he was an immensely influential figure in the Victorian scientific establishment in Britain to the degree that his views on criminality inevitably carried considerable weight. No doubt inspired by his own pedigree as a member of the influential Darwin-Wedgwood family, he published his most famous work, *Hereditary Genius*, in 1869.[53] He was influenced by Darwin's work and spent much of his working life exploring the statistics of heredity in human populations, being particularly interested in whether intelligence was inherited. He was the originator of the science of eugenics, or selective human breeding to improve racial stock. He contributed to the development of fingerprinting and had a significant role in promoting its adoption for criminal justice purposes in Britain, as the following chapter considers.[54] His other major contribution to scientific criminology was his composite photography.[55]

Photographic images of prisoners had been in use since at least 1850 in Britain with a photograph added to the registration of each prisoner following legislation in 1871.[56] Edward Du Cane, appointed to the chair of the Prison Commission in 1877, contacted Galton to help establish, through the medium of photography, whether there was a 'criminal type' with typical features associated with particular crimes. Photography was widely taken to provide pictorial realism in that period, yet it was circumscribed by various conventions which had to be understood in the production of 'ideal pictures'.[57] Photography, as part of the armoury of scientific objectivity, had to construct its meanings and had to convey signals as to how photographs were to be read.[58]

Galton was driven by his interest in the differences between races. Given the fairly small differences in the measurements of features between one individual and another, he did not feel that it was realistic to apply statistical methods to such small differences in physiognomy.[59] Instead, he followed the lead of natural historian and champion of Darwinian evolution, T. H. Huxley, in the production of photographs of typical members of a given race by contacting the Colonial Office in 1869 with a plan to photograph representative members of races in the British Empire.[60] While sorting through archetypal physiognomic characteristics he came up with the idea of superimposing photos to make a composite picture. For Galton, composite photography was a form of 'pictorial statistics' 'the purpose of which was to create a *type* of person'.[61] He saw himself as extending Quetelet's work – where Quetelet gave a statistical picture of *average man* in outline, Galton was able to produce a photographic picture of *average man* and then, by extension, a picture of *criminal man*. Galton was trying to convert the bell curve of the normal distribution into a picture of the face of a composite person, keeping the picture sharp in the centre of the face and leaving blurring to the edges of the composites rather than showing the blurring across the whole photo, which would naturally have occurred in the composite photos. The bell curve 'now wore a human face . . . Only an imagination that wanted to *see* a visual analogue of the binomial curve would make this mistake, finding the type at the center and the idiosyncratic and individual at the outer periphery'.[62]

The emphasis was on capturing a *typical* image of a criminal so an actual criminal could be identified prior to committing a crime. However, the images produced were bland and nondescript, hardly an advertisement for an exact science of pictorial statistics. Although he developed this work in terms of its criminological applications, Galton believed it was applicable to producing typical pictures of different races and for animal breeding, so he produced composite pictures for eugenic purposes. As Sekula notes, the images he produced in his *Inquiry into Human Faculty* 'amounts to an illustrated lecture on eugenics'.[63] In this work he produced a composite of six portraits of Alexander the Great and composite photos of ancient Greeks and Romans, 'seeking in the blurred "likenesses" the vanished physiognomy of a higher race'.[64] Galton's composite photographs enjoyed wide favour until the beginning of the First World War.[65] Indeed, they could be used to make a variety of points which Galton had not originally intended, including arguments for 'nurture' over 'nature', but as photography became more widespread in the twentieth century, excessive claims for truth began to diminish.

Galton's composite photography and Bertillon's anthropometric system for criminal identification (discussed in the following chapter) are both examples of 'a persistent, if often implicit, sentiment within much criminal justice thinking; namely, by identifying individual criminals, or criminal types, justice systems effectively address the problem of crime'.[66] Instead, we can understand the politics of seeing the identification of criminals 'not as a *discovery* but a *creation*'.[67]

However, Bertillon's and Galton's technologies for criminal identification were based on different scientific paradigms.[68] Bertillon's approach involved measuring and recording details, looking for ways of uniquely identifying the individual. Galton's involved technical means of identifying typical criminals or *criminal types*, part of 'the search for a way to identify criminal potentiality by finding some biological marker of latent criminality'.[69] So, whereas Bertillon's emphasis was the individual, Galton's was the type. Inheritance of physical, intellectual and moral faculties was key for Galton, and he wished to understand correlations between physical and mental characteristics amongst convicted criminals. His work thereby inherited concepts from the earlier, and by then largely discredited, sciences of phrenology and physiognomy. Unlike Bertillon, he was not interested in the individual, regarding small physical differences between individuals to be too numerous and usually too small to make meaningful comparisons; rather, it was the business of averaging via photography which was held to provide a reliable means of characterizing the whole class. Pavlich makes the important point that Galton's approach was tautological.[70] Rather than investigating *whether* there were habitual criminals, he started with those who had already been labelled thus within the criminal justice system and then set out to discover defining characteristics. If it was tautological, it was also not sociological. Galton, with his emphasis on hereditary intelligence, neither looked at the behaviour of criminals nor their social settings to understand their behaviour.

Although Bertillon's anthropometry and Galton's composite photographs embodied different scientific and criminological paradigms, they were both part of the civilizing and controlling project. This sought to identify the 'other', whether the other was an individual criminal or a 'criminal' race understood against the cultivated European ideal. The tools of measurement were brought to the project, including the tool of photography, where 'photographic portraiture had been formalized, made accessory to the state's and the law's need for systems of identification'.[71]

Lombroso and the born criminal

Cesare Lombroso (1835–1909), the founder of the Italian school of criminology, described *Criminal Man* by means of an anthropological approach.[72] Lombroso was fascinated by skulls and their shape; indeed, he acquired a significant collection of skulls, even leaving his own skull to be displayed in his museum after his death.[73] Against a belief in evolution and the privileged position of the scientist in the business of controlling and civilizing the criminal classes, he measured and categorized in prisons and clinics using an array of technical devices: 'the volumetric glove, pelvimeter, craniograph anfossi'.[74] Lombroso embraced the gadgetry of his subject, thereby bringing to criminology the combination of technology and science, an embodiment of the term 'technoscience' which the following chapter explores.

Lombroso's best-known work, *Criminal Man*, spoke of degeneration and decline.[75] Criminal 'types' were to be literally read from the body. As a medical doctor from Northern Italy, he was one of many who made a link between criminal and racial types to promulgate distinctly racist views. He regarded the 'atavism' that he identified in the people of Southern Italy to be related to the return to primitive stages also found in the criminal type. Galton's composite photography offered the visual promise of discovering criminal types and could be used to discover the state of a race, paving the way for the science of eugenics. However, for Lombroso it was the measurement of skulls and the categorization of markers of the criminal which offered a possibility of visually uncovering atavism, degeneracy and criminality against a backdrop of evolution.[76]

Lombroso's characterization of the *born criminal* became popular in the USA around the turn of the century.[77] Its promise of a scientific approach to the causes of crime held considerable appeal. As his work was not translated into English until 1927, American and British criminologists knew his work mainly through secondary sources.[78] Indeed, Havelock Ellis's book, *The Criminal* (1890), was a major source of information on Lombroso's work for English-speaking audiences.[79] Some American criminal anthropologists believed in the identification of a criminal class, where the *born criminal* was held to depart the farthest from normality.[80] They emphasized weak intelligence as well as weak moral understanding. Lombroso did not stress eugenic solutions to the problem of born criminals;

however, there was some enthusiasm for life imprisonment, even preventing marriage, to stop criminals breeding. As Rafter notes, there was no necessary connection between Lombroso's approach and eugenics; nevertheless, they became linked and a connection was made with a perceived degeneracy of the population, a concern in the USA and in the UK.[81] Lest we imagine that eugenic approaches in the twentieth century were largely confined to the atrocities of Nazi Germany, we should note that several US states had eugenic institutions which survived into the 1960s.[82] If the USA wishes to forget former enthusiasm for eugenics, so, too, does the UK. Eugenics still received backing in powerful places in the UK into the 1940s from all parts of the political spectrum, including support from prominent figures such as George Bernard Shaw, William Beveridge and John Maynard Keynes.[83]

Although it appeared to offer an evolutionary stance towards the criminal and an attempt to identify *criminal man* through biometric means, we should note that Lombroso's research was far from popular in Britain. Most European scholars rejected his scheme within thirty years.[84] Nevertheless, it offered a demonstration that a scientific approach using appropriate technologies could be used in identifying criminals, and that criminology needed to attend to the factors that divide criminals from ordinary people, even if it was hardly convincing to the British in its detail. His influence in Britain can be seen in terms of 'a "soft" adaptation of Lombroso's vision, stripping his thesis of all its particulars, but leaving in place the focus on the criminal . . . the gradual "medicalization" of the criminological endeavour'.[85]

Hence, although Lombroso's work, from the 1870s onwards, contributed to debates on the possibility of a scientific criminology, it did not greatly influence criminological approaches in the UK.[86] Instead, British scientific criminology derived from 'a distinctive, indigenous tradition of applied medico-legal science which was sponsored by the penal and psychiatric establishments, and it was this tradition which formed the theoretical and professional space within which "criminological science" was first developed in this country'.[87] Lombroso's lack of empirical knowledge made his work overly theoretical and abstract for the British penological establishment, who favoured a therapeutic approach based on a spectrum of psychiatric disorders and a knowledge of the criminal justice and prison systems, supported by the medical establishment.[88] Some who wrote about criminology in Britain were medically trained doctors with positions in the prison service, e.g., prison medical officers and the Medical Commissioner of Prisons.[89] Goring's (1913) study, *The English Convict: A Statistical Study*, continued this practical tradition.[90] His study of 3000 research subjects, although finding criminals on average shorter, lighter and less intelligent than non-criminals, found no evidence of Lombroso's criminal types. *The English Convict* was more popular in the USA than in the UK, possibly because its focus was on statistical rather than clinical methods for obtaining criminological knowledge. Nevertheless, the work formed a bridge between older and more modern conceptions of scientific criminology. It engaged with long-standing ideas about criminal types, yet viewed criminality as part of the spectrum of normal behaviour rather than characterizing a separate criminal type or types, at the same time offering statistical analyses which suggested new approaches to the study of criminals.

Davie argues that the opposition of the Lombrosian school and the British pragmatic school was not as extreme as it has generally been painted; rather, the British stuck the bits that they liked onto their therapeutic approach towards criminality.[91] The issue was not so much that the British criminological establishment was opposed to the idea of the *born criminal*; rather, it was opposed to the idea that physiognomy and expression could predict criminality through identification of external signs. The idea of a 'criminal type' seemed to deny the complexity of individual cases and the possibility of treatment and redemption. As James Devon, medical officer for Glasgow prison, noted, speaking of the eminent participants at the 1896 Geneva Congress on Criminology, 'Anyone who glances at the illustrated papers will see for himself as many villainous-looking faces among notable people, even among able people as he will find in prison.'[92]

Hans Gross and criminalistics

My intention is not to find an unbroken path from Lombroso and Galton to twentieth-century forensic scientists. The influence was more subtle and less direct, forming a backcloth against which scientific analyses of crime were conceived; often an author's work would be known through translations and description, as with Lombroso's and Gross's research. As already noted, Pavlich argues that Galton's approach was tautological.[93] But if he was guilty of tautology, then others must share the burden of tautological guilt. The whole business of classifying *born criminals*, *habitual criminals*, *criminal man* or *criminal classes* assumed the prior existence of the habitual criminal and said little, if anything, of how the criminal could be linked to an individual crime. Scientific criminology gave no clues to the detective or police officer as to how to connect the criminal to the crime scene.

As Vyleta notes, historical attention has centred on mainstream criminology and medical understandings of criminality to the neglect of criminalistic thinking on crime: 'which combined interest in policing and identification technology with a trenchant critique of the mainstream criminological endeavour'.[94] An alternative approach which acknowledged the importance of the crime scene and emphasized ways to manage and analyse its contents, which increasingly came under the banner of 'scientific detection and policing', heralded the development of forensic science proper. Increasingly, 'middle ranking' scientists and police detectives were dealing with the minutiae of analysing trace evidence rather than making more abstract pronouncements on the features of *criminal man*. Although attempts were made to make policing more 'scientific' by introducing criminological ideas in the education of criminal justice practitioners, the methods of Hans Gross were much more influential in the development of scientific detection and policing and, ultimately, the creation of forensic science.[95] Burney and Pemberton argue that opening up the distinction between criminalistics and criminology gives opportunities for new conceptualizations, with the role of the investigating officer and the crime scene taking centre stage.[96]

Although criminalistics can be characterized as representing something of a break, it shared common underpinnings with scientific criminology. Measuring and mapping were important in developing the language of a scientific approach. The belief in the authority of appropriate witnesses and experts as the conduit of scientific objectivity was also part of the story. Finally, evolution played a part as the justification for 'othering' and demonizing criminal races and setting criminal classes apart. The virtue of Gross's approach was that it offered the possibility of identifying the individual criminal and linking him or her to the scene of the crime; therefore, it was an idea which came with instructions on how to apply it, which scientific criminology did not. His genius was not so much in terms of original scientific techniques, although he did expend much effort on maintaining his understanding of science, particularly as his career developed in the academy. The skill of Gross lay, rather, in understanding how these techniques might be applied to the detection of crime and how the crime could be linked to the criminal rather than by searching for physical markers of criminality on the individual criminal.

Although often hailed as the father of criminalistics, he was not unique in his attention to the crime scene.[97] In the UK, the development of policing techniques and the rise of the detective role were beginning to cast the spotlight on criminal detection and the management of the crime scene itself in the twentieth century.[98] Small wonder that a similar approach was eventually taken up in the UK once the higher echelons of Scotland Yard realized it could be used to support new thinking in scientific detection in the 1930s. Although he was not necessarily the first, nor the only one, to describe a system of management of the crime scene around the turn of the century, nevertheless, his *Systeme der Kriminalistik* or *System of Criminalistics* was remarkably sophisticated in its understanding of the whole ambit of the crime.[99]

Gross was interested in measurement, and he was interested in criminal psychology underlining the links between his subject matter and scientific criminology. He was also interested in the psychology of witnessing, recognizing that the memories of witnesses, even when corroborated by other witnesses, could be unreliable.[100] Lombroso was influenced by his medical and military background, whereas Galton's interest in hereditary factors and eugenics inspired his search for physical markers of the criminal. Gross's professional position as an examining magistrate put him in a unique position to gain an encyclopaedic knowledge of crime. As Becker notes, the inquisitorial system placed the burden on investigating magistrates not only for gathering evidence from witnesses, but also for describing the crime scene objectively.[101] His ability to encompass the whole range of criminal activity, including control and management of the crime scene, placed him in a unique position to understand the scene of the crime in relation to its scientific possibilities.

Initially qualifying as a lawyer in Graz, he became an examining justice in the Upper Styria district and later a public prosecutor in Graz.[102] Austrian police forces were composed of former soldiers. 'Without the aid of any technical apparatus they knew very well how to keep up peace and order by merely applying mother-wit and common sense supported by the authority of the Imperial uniform.'[103]

When Gross took up his legal appointment in the 1870s, he was shocked to realize that his university training in law was of so little use, as it said nothing about establishing the facts of a crime. He saw only too well the paucity of support for those whose jobs it was to prevent and detect crime. His purview included a vast array of criminal acts, including varieties of fraud, arson and murder. Necessity was, if not the mother, then at least the midwife of Gross's invention. Given that there were no trained criminal investigators when he set out on his career, it was the job of the examining justice to establish the facts of the crime. With an eye for detail, a vast array of experience and a burgeoning interest in the sciences, he was led to develop the discipline of criminalistics. His challenge was to reconcile traditional wisdom with the development of modern scientific criminological knowledge. 'Gross tried to solve this problem . . . by transforming the claims of truth and objectivity by the investigating magistrate into mechanical reproductions of material evidence.'[104] He operated within an inquisitorial system of investigation where the examining magistrate had to reconstruct the truth of criminal events, a truth which could be hindered by the deceptions of criminals and the memory or unreliability of witnesses, hence his interest in criminal psychology. But the truth could also be frustrated by the magistrate's inability to gather and present evidence to an appropriate level of objectivity. Magistrates had an 'obligation to represent objectively all information about the crime. However, objectivity was not understood as the mechanical reproduction of depositions but rather as the skilful, reasonable recreation and evaluation of events'.[105]

Grassberger, Gross's biographer, argues that this provided the impetus for his research and also fuelled his desire to publish his results to make these available to others in the criminal justice system.[106] He devoured a vast amount of scientific literature on physics, chemistry and medicine, acquainting himself with microscopy, photography and x-rays in order to establish how these sciences could be employed in the detection of crime. His interest in scientific and technical details was stimulated by his understanding of a criminal case as a scientific problem to be solved.[107] He knew that the abstruse scientific descriptions in the literature were difficult for the student of law, hence his painstaking description in the *Handbuch*. Consulting the appropriate expert witness was a fundamental part of his criminalistic scheme.

Gross became professor of criminal law and justice administration in Czernowitz, moving later to the University of Prague, then Graz in 1905, where he opened a criminalistic institute for research and teaching in 1912.[108] His students were exposed to all aspects of criminalistics and police science. He built up a collection of criminological objects; his criminological museum can still be visited today. His legendary attention to the detail of the crime scene, his widely read publications which ran to several English-language editions and his initiation of a teaching institute for criminalistics may make Hans Gross appear to be a paragon of modernistic forensic virtue. But it is important to recall that his view of criminals was infused with a similar racism and imperialism to that of Lombroso and Galton, with a large heaped spoonful of paranoia added to the mixture.[109] He was just as much of his time as were contemporary scientific criminologists. He believed that understanding habits of the criminals, their means of working and communicating, including criminals' secret means of sharing information, were as important as the scientific aspects of criminalistics. He was particularly interested in professional criminals, or *Gauner*, and the tricks and deceptions of *Gauner*, namely *Gaunerpraktiken*.[110] These included deceptions such as fraud, swindling and embezzlement, disguises and impersonations. He projected the character of *Gauner* onto gypsies as an ethnic type. As Becker and Vyleta note, his approach to gypsies/*Gauner* was particularly paranoid and authoritarian. He was influenced by ideas on evolution, understanding the supposed habits of gypsies in terms of a kind of cultural evolution rather than the breeding of criminality by physical evolution favoured in scientific criminology.[111]

Hence, Gross's position within a particular type of inquisitorial system where he was responsible for managing the crime scene, gathering and preserving evidence, arranging the prosecution and presiding over the court case was crucial to the development of his criminalistic approach.[112] Epistemological matters were heavily influenced by institutional arrangements; indeed, they were inseparable. In the Anglo-American system, the activities encompassed by his wide-ranging role were divided up amongst different professionals in the adversarial legal system. Arguably the adversarial legal system in the late nineteenth century did not offer the same opportunity for a senior official to develop the skills and interests to encompass the range of forensic interests demonstrated by Gross, as the responsibility for investigating the crime and for presenting prosecution and defence evidence belonged to a number of individuals rather than to an individual examining magistrate. Indeed, as there was no exact role of examining magistrate, the English translations, which were clearly intended for a wide audience, used the term *investigating officer*, or *investigator* which, arguably, were not the same terms as *examining magistrate*.

Criminal Investigation **and colonial criminalistics**

Gross's *Handbuch für Untersuchungsrichter als System der Kriminalistik* or *Handbook for Examining Magistrates* was first published in 1893.[113] It ran to six further editions in German, acquiring the subtitle *System of Criminalistics* from its third edition in 1899 and with the final German-language edition published in 1943.[114] The first English translation and adaptation was published in 1906 as *Criminal Investigation: A Practical Handbook for Magistrates, Police Officers and Lawyers*.[115] *Textbook* was substituted for *Handbook* with the publication of the second edition in 1924 and it continued as *Textbook* in subsequent English editions.[116] The book ran to five English-language editions with the fifth edition published in 1962.[117] It is remarkable that it remained relevant for such a long time, but much of its longevity in English translation was due to the fact that it was significantly rewritten, augmented and adapted with successive editions.

As documentation of cultural change, the value of examining subsequent editions of a work has not always been appreciated. Certainly, in this case, there are some important insights about the development of crime scene investigation and the delineation of professional detective roles in the English-speaking world to be had from a brief consideration of the English editions of Gross's *Handbuch* which went far beyond his original writing. The changes to the various editions act as a mirror reflecting developments in the policing of the empire and policing at home. The English editions were translations *and* adaptations of Gross's book; hence, from the start his work was adapted to meet the requirements of particular and evolving audiences – they were not literal translations. They were rather a kind of palimpsest, where the original book, the 1904 edition of the *Handbuch*, was translated and adapted in 1906 then reworked and remade to suit different criminological purposes.

Drawing parallels with fingerprinting as a criminological technology that was developed in a colonial setting, it is notable that Gross's system also arrived in the UK via her most important colony.[118] John Adam (1850–1914), Barrister-at-Law, Crown and Public Prosecutor, Madras and his son, J. Collyer Adam (1879–1928), Barrister-at-Law, Advocate, High Court, Madras were the translators and editors of the first English edition, published in Madras and reprinted the following year in London, and Collyer Adam was editor of the second edition, which was published in London.[119] These were produced 'primarily for the benefit of Indian and colonial magistrates, lawyers and police officers'.[120] In the Madras Presidency, where the Adams worked, colonial policing was highly centralized, with the administrative role of magistrate-collector responsible for collecting revenue, administering the district courts and supervising the police.[121] Although responsible for police administration, this was not the same role as Gross's examining magistrate. Adam and Adam emphasized in their preface that their intention was to bring Gross's book up to date; however, their translation and adaptation was made of his 1904 edition and published in 1906! It is hard to see what could have become out of date in such a short space of time. Rather, it appears that they were referring to the reworking of Gross for the colonial criminal justice situation. Many of the illustrations and illustrative cases they incorporated reflect the colonial origins of the translation. Various drawings of material provided by the chief constable of Madras were supplied by the Madras government; a catalogue of weapons was provided by the Indian Museum, London; further illustrations were given by a Madras gunsmith and a Madras publisher; and botanical drawings were produced from specimens supplied by Lt Col Van Geyzel, the chemical examiner to the government of Madras.[122] The provenance of the illustrations and that they concentrated on weaponry and flora local to Madras serves to underscore the Adams's adaptation of Gross's work for colonial use.

However, apart from local illustrative examples, there were much more significant reasons why the first English editions of Gross should have originated in colonial India. These relate to major themes of Gross, the demonizing of wandering tribes and the related question of unreliability of witnesses, even so-called expert witnesses. Gross regarded gypsies as a criminal class. Although critical of Lombroso's Italian school of criminology, he was convinced that some criminals had their criminality biologically determined. He saw vagabonds and political revolutionaries as dangerous, advocating deportation and even castration for some criminals.[123] His campaign against gypsies was based on the idea that they were a criminal underclass, which he regarded as belonging to a separate evolutionary branch from respectable Europeans. It was not difficult to project his views on gypsies as *Gauner* onto nomadic Indian tribes, to make definite comparisons between the two. Adam and Adam, discussing nomadic Indian tribes, referred to the traditional 'gipsy industries' of mat and basket making as 'an easy cloak for more nefarious practices, as is the case with the European gipsy. . . A large part of the crime of the country is due to these people'.[124] One of the major concerns about crime in Madras centred on 'dacoity', or banditry, and nomadic groups were often held responsible.[125] Methods of reforming Indian wandering tribes were perceived to have failed by the British authorities who believed that nomadic tribes did not want to follow 'respectable' occupations and they 'constantly desert their settlements'.[126] Whereas only a few pages of the first edition of *Criminal Investigation* directly referred to nomadic Indian tribes, much of the book focused on superstitions, habits and activities of criminals and criminal groups, where the implication can be drawn that such considerations were to apply to local wandering tribes.

Adam and Adam's views were expressing contemporary beliefs on criminal tribes in India, which were interwoven with evolutionary views of race that extended ideas about habitual criminals. Scientific criminology had, of course, already thrown the spotlight on habitual criminals in the proceeding century.[127] They reworked Gross's views on European gypsies to reflect early twentieth-century colonial criminological orthodoxy on criminal tribes. The British regarded lower-caste nomadic people as criminal tribes, and indeed regarded the habitual criminal as a hereditary criminal in India.[128] Sedentary communities looked upon those who were nomadic and whose lives did not conform to traditional, settled agricultural roles with considerable suspicion.[129]

In the Madras Presidency, the Yerukulas were a significant nomadic group who became officially criminalized around the time when the Adams first published their work. The Yerukulas and other 'wandering tribes' of the Madras Presidency were declared criminal tribes in 1913 under a revised version (1911) of the Criminal Tribes Act (1871).[130] The new act applied to all of British India and provided for the 'reform' of a criminal tribe if it could be confined to a settlement to be kept under constant surveillance.[131] The Yerukulas were 'reformed' by the Salvation Army to work on its land and to provide cheap regular labour for a tobacco factory, thereby conforming to British economic policies designed to raise revenue, against which wandering people were regarded as a threat. The whole process, in the early years of the twentieth century (the Yerukulas were resettled in 1913), was so successful from the colonial point of view, that contemporary Yerukulas still believe that their forebears were criminals.[132] Booth-Tucker, head of the Salvation Army in India, noted that their work in Madras started late but quickly overtook other presidencies and provinces. 'Nowhere in India have the provisions of the CTA been so effectively taken advantage of as in Madras.'[133]

Hence, colonial laws marginalized their whole way of life, which was regarded as threatening, and colonial policies did much to destroy their traditional means of making a living. They were regarded as hereditary criminals passing on the criminal profession to their descendants. The British transported many of the prejudices they had about gypsies 'and seem to have simply superimposed some of these on the Indian counterparts'.[134] As Cole notes, 'Colonial jurists' and anthropologists' corruption of evolutionary theory in the service of British racism demanded the development of new methods of tracking, monitoring, and controlling suspect populations'.[135] Settlement was one method of control; surveillance based on Bertillon's methods and fingerprinting were, of course, other methods. Hence, Adam and Adam produced their translation and adaptation of Gross's *Handbuch* against a backcloth of the demonizing of nomadic tribes in India, where European distrust of gypsies could be superimposed on the caste system of India, and they were writing at the time when the Criminal Tribes Act was about to be enacted in the Madras Presidency with subsequent considerable success. So, the Adams's adaptation of Gross's *Handbuch* can be seen as another brick in the edifice of successful attempts to criminalize indigenous people under the British Raj.

The Adams translated the term *Untersuchungsrichter* in the original title of Gross's *Handbuch* as 'Investigating Officers' rather than 'Examining Magistrates', the latter being closer to the original meaning of the term. They explained that Investigating Officer was a 'compendious term to include all persons engaged in the investigation, official or non-official, of criminal cases'.[136] In India this included magistrates, police or interested parties and their legal advisers. Although Gross meant his book to be widely used in criminal justice applications, this signalled a difficulty in how his work was to be understood and adapted for a differently organized criminal justice system. The examining magistrate was the official who gathered all the data on the crime under the inquisitorial system in which Gross worked. Britain and her colonies operated under an adversarial system where the judicial and policing roles were quite different. Nevertheless, the Adams's term, 'Investigating Officers' was used in the third English edition, published in the UK in 1934; subsequent editions added the term 'Investigators'.[137]

The other major theme taken up and remade in the colonial setting was the question of reliability of witnesses, expert or otherwise. Unreliable witnesses were, of course, a problem everywhere. However, once again, the colonial situation was seen to present particular issues of unreliability; Adam and Adam argued that the 'value of the depositions of even a truthful witness is much over-rated'.[138] There were numerous errors in perception, and this led them to a typical 'silent witness' style of argument that the physical evidence was more reliable than human evidence. 'As the science of Criminal Investigation proceeds, oral testimony falls behind and the importance of realistic proof advances; "circumstances cannot lie", witnesses can and do.'[139]

In India, experts 'of the first rank' were rarely available.[140] To be sure, experts from official government organizations, including 'Chemical Examiners to Government, the Lecturers on Medical Jurisprudence at Headquarters, the Government Experts in Finger-prints' were people who knew how to give evidence properly. But this was not true of lower grades of hospital doctors and apothecaries, who might be near the beginning of their careers. Adam and Adam argued that this sort of medical person 'can seldom be called an expert at all'.[141] They were particularly negative about the situation in the Madras Presidency where hospital assistants were asked to undertake post-mortems and to report and give evidence in murder trials.[142]

> The ordinary official of this grade does so without the slightest hesitation, makes statements and draws conclusions with the utmost rashness declining to admit his inability to reply to a question a far more experienced man would shrink from answering, and entirely forgetting that the life of a fellow human being may be dependent on his words. This is no random assertion: the records of our Criminal Courts amply bear it out, and high medical authority could be quoted for the opinion that no non-gazetted medical officer *i.e.* no one below the rank of Assistant Surgeon, should be permitted to conduct a *post-mortem* examination in a suspicious case and make a report thereon.[143]

The Adams regarded medical men in general as poor at giving evidence and holding their own under cross-examination in court.

> They may know their trade thoroughly and be able to give most useful information, but are incapable in a criminal matter of applying their knowledge and answering questions put to them with any precision. Without the slightest intention of frustrating the ends of justice, they fall victims to the wiles of the expert cross-examiner, losing both their heads and their tempers.[144]

Even if they were expert in their discipline, this did not mean that they were able to draw up a report for a criminal court and answer questions in the right way. This echoed Walls's point that medical practitioners did not always make good expert witnesses because the way they thought and expressed opinions for medical decisions did not match the kind of expert opinions that are required in court.[145] If so-called experts had trouble expressing themselves effectively in court, what hope would there be for the local hunter or agricultural worker? This was all the more reason why the investigating officer was needed to take charge and extract the knowledge required from specialists.

In itself, the problems with expert witnessing did not look very different from the situation in England. However, Adam and Adam were making a clear distinction between experts to be trusted, namely high-up officials in laboratories and government departments, and lower-graded medical doctors and local people based on a strictly applied hierarchy. The ability to make these clear distinctions was considerably facilitated by the highly structured Indian system of distinguishing between gazetted and non-gazetted officials. Effectively, they were distinguishing between the (mainly) Europeans who would largely have been in senior positions and Indian people, trusting the former but not the latter. For instance, the pages of the *Institute of Chemistry's List of Official Chemical Appointments in Great Britain, Ireland, in India and the Colonies* for 1906 for India confirms that, in chemistry at least, senior positions were held by those with European, usually British, names.[146] Indian names, where they appear on the list, were usually in 'Assistant' roles. For instance, the chemical examiner in Madras in 1906, a contributor of material to Adam and Adam's book, was Lt Col John Lawrence Van Geyzel who had obtained his medical degree in Aberdeen.

Gross saw the ideal witness as someone with sufficient maturity, reliability and interpretive powers. Such a person was usually in an official role or was a scientific expert witness and so would have the necessary authority to offer objectivity. However, the expert witness was not usually at the crime scene when the crime was committed. Because it was not possible to trust the perception of ordinary witnesses, other information about the crime had to be collected and 'questioned' using the appropriate technological methods.

> The graphs, charts, numbers, and microscopic images . . . had to be read by expert witnesses . . . . Modern research therefore put an end to the reliance on language as a privileged approach to the truth about criminal acts . . . new technology promised police and legal experts an 'objective,' depersonalized reconstruction of crime scenes.[147]

The objectivity and reliability of such processes were taken for granted. Gross's achievement can be seen in terms of a move away from trusting the reliability of the ordinary witness towards trusting the reliability of the 'silent witness' of material evidence as interpreted by the scientific expert witness and criminal justice official, who were people of appropriate experience and authority such that their objectivity could be guaranteed.

These were the twentieth-century versions of Boyle's virtual witnesses, but Enlightenment views of the inherent rationality of human beings had disappeared; only certain witnesses were to be seen as reliable. The disciplinary society of the scientific criminologists, coupled with Gross's strict management of the crime scene where only objective evidence, the silent witness, could be trusted and where there were strict limits both to the trustworthiness of criminals and criminal groups and also to the reliability of experts signalled a move away from trusting inherent human rationality towards identifying the role of the examining magistrate or investigating officer as the disciplinary official with the appropriate experience and position to take charge of the crime scene and call in appropriate scientific expertise. The highly structured and hierarchical nature of Indian colonial professional life made it relatively simple for the Adams to transpose Gross's strict views on the value of expert witnesses onto colonial criminal justice. In summary, the first two English editions of *Criminal Investigation*, translated demonizing gypsies into criminalizing Indian wandering tribes and Gross's mistrust of witness testimony was superimposed on the colonial hierarchy of expert positions.[148]

Remaking *Criminal Investigation* for 'scientific aids' in Britain

Only with the publication of a third edition in 1934 was the English translation definitively moved to British soil and, from that time, *Criminal Investigation* incorporated British examples and references, including references to a number of independent and Home Office forensic scientists.[149] This edition, and subsequent editions, was badged as a *Practical Textbook* rather than a *Practical Handbook*. Although this may seem to be a small change, it appeared to signal that the later, i.e., *British*, rather than the original colonial, edition was to be seen as instructional rather than necessarily, to be used in the field by the investigating officer. In the 1934 edition, the chapter on 'Wandering Tribes' discussed European gypsies with no mention of the criminal tribes of India; indeed, the Indian examples which the Adams used in their earlier editions were removed. The editor of the third edition was Norman Kendal, CBE, Assistant Commissioner of Criminal Investigation Department, Metropolitan Police. On the advice of a number of British scientific and medical experts, Kendal claimed that he had revised the work, thanking the editor of the *Police Journal*

> for permission to make use of a large number of articles which have appeared from time to time in that paper, particularly articles on Firearms, by Major Hugh Pollard, Wounds Caused by Firearms by Professor Sidney [sic] Smith, The Fluorescence Test and Ultra-Violet Light by Herman Goodman and Forensic Chemistry by Mr. Henry Rhodes.[150]

Kendal also acknowledged the help of Home Office analyst Roche Lynch on blood tests and the well-known gunsmith and ballistics expert, Robert Churchill, whom we met in Chapter 1. As we shall see in Chapter 4, the *Police Journal* was one of the leading lights in promoting scientific aids in policing, publishing many articles on such matters in the 1930s. It is not unexpected that, in the hands of an editor versed in modern British scientific policing, *Criminal Investigation* would be moved away from colonial interest and moulded into a text for the 'scientific aids' movement.

The fourth edition of *Criminal Investigation* was published in 1949, after the war, with R. M. Howe, Barrister and Assistant Commissioner of the CID at Scotland Yard, as editor. This was a much slimmer volume, and Howe claimed that '[i]n spite of the drastic cuts and alterations I have felt it necessary to make, this book still remains the work of the great Dr Hans Gross, who was years ahead of his time'.[151] Not surprisingly, as this was published not long after the end of World War II, the chapter on 'wandering tribes' was removed (and, of course, did not return in the final edition) and a chapter on road accidents was included instead. The central character became the 'investigator' rather than the 'investigating officer'; this signalled the consolidation of roles in detection and forensic science. Whereas 'investigating officer' appeared to refer to a detective and therefore a police officer, 'investigator' was a more fluidly defined role, not necessarily mapping onto one police official.

However, the hacking, chopping and remaking of Gross's original work (or rather his 1904 version), although still attaching the name of Gross to it and naming Adam and Adam as editors, did not pass without comment in all quarters. G. W. Wilton, a retired sheriff in the Scottish legal system, was highly critical of Howe's 1949 edition.[152] As Wilton pointed out, four more German-language editions of Gross's *Handbuch* had appeared between 1908 and 1949, all produced after the 1904 edition on which the English-language works were based. 'No research by any Adaptor into his writings after 1904 has apparently been made.' He described Howe's adaption as a 'hash' of the three earlier English editions with 'much material [is] thrown in . . . the Gross of 1904 . . . is hardly discernible'.[153] Howe did not appear to have consulted any of the later German-language editions and had assumed that there were no copyright issues. Indeed, Wilton concluded that it was doubtful that Howe's book was fair use of any copyright granted by Gross to the Adams. None of these individuals was still alive to dispute copyright. Indeed, Gross completely rewrote his 1914 edition. Wilton concluded that the assertion that this remained the work of Gross was very dubious and it was problematic as a work of criminal investigation because of its omissions.[154]

The fifth and final edition was edited by Richard Leofric Jackson, CBE, Barrister-at-Law and Assistant Commissioner of the Criminal Investigation Department, New Scotland Yard, in 1962.[155] Jackson added a further chapter on international crime on the basis that crime was no longer local; it had become organized and international. Jackson claimed that the achievement of Gross was to have understood the basic processes required which could then be augmented by burgeoning expertise in the sciences.

> The changes I have described make it necessary for the Investigator to employ that services of experts in an ever-increasing variety of fields of knowledge but the general guidance given by Gross to the Investigator remains as important as ever.[156]

Jackson acknowledged the help of a number of senior police officials, notably, L. C. Nickolls, Director of the Metropolitan Police Laboratory, and J. B. Firth, by then, former Director of the North Western Forensic Science Laboratory.[157] Once again, the scope of *Criminal Investigation* had grown beyond British shores to include international crime but to drop colonial crime. *Criminal Investigation* had travelled around the world in sixty years.

The vision of Gross's work can be seen in the way it mapped out a series of operations for the investigating officer to undertake when called to the scene of a crime and, importantly, when to call in the services of an expert. The distinction between the roles of scientist and police officer, and what were the proper and separate duties of both, were extremely important and the subject of negotiation. Gross was one of the first to make a clear professional delineation between investigating police officer and scientific expert, although he was certainly not the only voice to speak of clear professional demarcations.[158] Collyer Adam's introduction to the second English edition, originally published in 1924 and reprinted in the 1962 edition, emphasized this distinction.[159] Adam argued that the stage of the inquiry where the expert is appointed depends on the knowledge of the investigating officer; one who is well versed in the capabilities of scientific analysis would know when to call in an expert, otherwise, there would be little chance that the evidence would be placed in expert hands. '[T]here is a vast gulf between permitting an Investigating Officer to undertake work beyond his sphere and instructing him when to recognize *when he ought to resort to experts, what experts should be chosen, and what questions must be submitted to them.*'[160] Investigating officers were expected to act as an expert when little expertise was required. Examples include 'falsification of documents, observation of footprints, the deciphering of writing, questions concerning superstition etc.'.[161] It is interesting that, with the exception of 'superstition' and footprint observation, these areas would eventually become the remit of the specially trained expert. This serves to emphasize the contingent nature of expertise and that division of labour between scientist and police official was continually negotiated; what was regarded as specialist knowledge and what was not varied at different times. Some things later became the subject of expert scrutiny, e.g., shoeprint analysis. However, document analysis was by this time coming under the purview of expert analysis.[162] Indeed, as Chapter 1 describes, it was a subject that was invented for forensic purposes rather than being a pre-existing science applied to criminal detection.[163] Who was to be regarded as an appropriate specialist for a given specialist examination developed considerably over the period while Gross's English edition was in print, and this was reflected in the expansion of appropriate sections of the book through subsequent editions.

An investigating officer would have had to assume the role of the expert in the case of an emergency where there would be no time to call in further expertise and an arrest had to be made on the spot because the investigating officer needed to apply his own knowledge to go beyond written procedures to get the right expert to answer the required questions. In that regard, the work was to be 'a book of "First Aid"'.[164]

Contemporary understanding of the term criminal psychology focuses on the psychology of the criminal with much interest in criminal profiling. However, Jackson's 1962 edition of *Criminal Investigation*, repeating the words of Adam written in 1924, gave a much wider definition of the term. He emphasized that the character and behaviour of the witness was as much a part of criminal psychology.

> [T]he character of the criminal and . . . the weight to be attached to the testimony of the witnesses, the consideration of errors in observation and deduction, to which judge, jury, and all who have to deal with crime, are exposed – these things are part and parcel of our subject.[165]

In Chapter 1 I argued that successive editors of a canonical work of medical jurisprudence, *Taylor's Principles*, gained prestige by their association with a venerable work over its consecutive editions.[166] Meanwhile the book was refashioned whilst retaining the intellectual authority of the original author no matter how much later editions strayed from the original. A similar process took place with successive English-language editions of *Criminal Investigation* where successive editors were at pains to emphasize that their editions were produced in the spirit of Gross, with recognizably Grossian ideas, despite, if Wilton is to be believed, hacking about the 'authorized' first edition by Adam and Adam and despite ignoring Gross's own much revised later edition of 1914.[167]

Alfred Lucas, *Forensic Chemistry and Scientific Criminal Investigation*

The English editions of *Criminal Investigation* were not the only way in which Gross's ideas, and similar views on crime scene management, reached British shores. Arguably, the successful marriage of forensic science with strict management of the crime scene in the style of Gross (and citing Gross as a reference point) was first achieved by a British analytical chemist working in Egypt, Alfred Lucas (1867–1945), in his *Forensic Chemistry and Scientific Criminal Investigation* which ran to five editions between 1920 and 1945.[168] Lucas, when he is remembered at all, is known for his work on ancient Egyptian materials and on archaeological conservation; he worked for nine seasons alongside Howard Carter on the Tutankhamun excavation, and is widely credited with preserving over 90 per cent of the tomb's contents intact.[169] Until his retirement in 1923, he was Director of the Government Chemical Laboratory in Cairo, in which role as an expert analytical chemist, he was often called upon by the Egyptian criminal justice authorities to advise on forensic matters, earning the soubriquet 'Egypt's Sherlock Holmes'.[170] Colonial government chemical laboratories in the early years of the twentieth century undertook a wide variety of analyses, civil and criminal, thus providing fertile training grounds for early forensic scientists. C. T. Symons, who enjoyed a brief career in the UK as forensic science adviser until his early death in 1937, had been in the equivalent role in Ceylon prior to his return to Britain.[171]

Based on his experiences in Egypt, and using many local examples, Lucas produced his first two editions while still at the Government Laboratory, continuing to revise and expand the book in subsequent editions. He was often called as an expert witness right through his life. His work is quite different from the only other substantial English-language work of forensic chemistry at the time, William Jago's *Forensic Chemistry and Chemical Evidence*, published in 1909 and with an emphasis on the legal aspects of evidence.[172] Instead, Lucas's *Forensic Chemistry* emphasized the management of the crime scene and how that related to the kind of assistance an expert analytical chemist could provide to the police.

Living and working in Egypt from the late nineteenth century until his death in 1945, Lucas published most of his many papers on Egyptian materials and archaeology in English, but some he wrote for Francophone Egyptian academic periodicals.[173] An educated British man in Egypt over that period could be expected to achieve fluency in French, and it is no surprise that he drew upon a number of books in French on the subject of forensic chemistry and crime scene management in his *Forensic Chemistry and Scientific Criminal Investigation.*[174] The work of Locard in his forensic laboratory in Lyons was well known – arguably he achieved an even more legendary status than Gross in forensic circles.[175] However, it was work of R. A. Reiss, first at Lausanne then later working in Belgrade, that Lucas successfully brought to readers of his book.[176] Like Gross, Reiss had much to say about the character of professional criminals, gangs, tricks and language, but he also laid out highly detailed descriptions of what to look for at crime scenes and different kinds of thefts and murders. This was another significant contemporary work of scientific criminal investigation and crime scene management interpreted for British and colonial readers.

The direct influence of Lucas's book on British scientific detection and forensic science is difficult to judge. Nevertheless, it was very widely referenced in the UK. Almost everyone writing on forensic scientific matters or scientific crime management and detection in the UK from the mid-1920s for the next thirty and more years cited Lucas. Even H. J. Walls in his textbook on *Forensic Science* published in 1968 still referred to *Forensic Chemistry* as a standard reference work on the subject, although by then somewhat out of date.[177]

Although *Forensic Chemistry* was not a work intended for popular consumption in the way that Charles Ainsworth Mitchell's science and crime books clearly were, Lucas came to the notice of a more popular British audience both through his links with a number of writers of crime fiction, as Chapter 6 describes, and through the writing of Nigel Morland, who was an active correspondent of Lucas's. Morland was a journalist turned crime fiction writer who produced a huge opus of detective fiction in the middle years of the twentieth century, some of it in the 'scientific detection' mould, under a variety of pseudonyms and all now largely forgotten. However, he also edited a criminological journal and, although apparently without scientific training himself, produced a number of popular science and crime books.[178] Morland was a well-known figure in the penumbra of forensic science; he was active in the Forensic Science Society in the 1960s and regularly reviewed books for their journal.[179] He actively promoted Lucas's work through his popular books from the 1930s to the 1970s, acknowledging his help and dedicating *The Conquest of Crime* to him.[180] In particular, he quoted Lucas's definition of forensic chemistry with its delineation from forensic medicine, unchanged through the years, and reminded popular audiences and specialist audiences that Lucas's experiments with quicklime demonstrated its preservative properties which were significant in the Crippen case (described in the following chapter).[181] Lucas's marriage of crime scene investigation and forensic science coupled with his reference to Gross's work and other continental forensic writers was thus emphasized for specialist and popular audiences in the UK through Morland's efforts in the middle years of the twentieth century.

Conclusion

A number of themes emerge from the discussion of the development of scientific criminology in the UK and the influence of criminalistic approaches, particularly that of Hans Gross, which arrived in the UK via colonial translations. Both approaches enrolled science and scientific ways of thinking, especially measurement and the concept of an expert observer able to judge objective physical evidence, to give authority to their views on criminals and crime. In maintaining order in society, criminal classes could be cast as the 'other'. Nowhere is this more apparent than in Gross's attempts to demonize gypsies whom he regarded as *Gauner* with their supposed tricks and deceptions, and following Gross, Adam and Adam's identification of criminal tribes in their adaptation of his work.

Gross's mythic status was reinforced by the remaking of his work through its English-language editions over more than fifty years, from a colonial setting then into Scotland Yard and British scientific detection. Scientific criminology lent authority to the medical and scientific gaze. The criminalistic approach also added further emphasis to the authority of the expert. In the two first English editions of Gross's *Handbuch*, the emphasis on the expert was transposed onto colonial hierarchies of medical and scientific experts.

Scientific criminology and criminalistic approaches, encompassing a wide field of authors, created the criminal, albeit in different ways. Although different in emphasis, they were both instrumental in reinforcing the authority of the scientific expert and in constructing a 'scientific' approach towards criminal and crime detection that underpinned the development of forensic science in the twentieth century. However, there is a wider story to be told about how these approaches were incorporated into scientific detection and how technologies, including, significantly, information and communications technologies, were enrolled into the business of scientific detection of crime. Chapter 3 turns to the question of technoscience in scientific policing, using the Crippen case as an exemplar.

Notes

1 Not all forensic science is directed towards the attempt to link a potential criminal to the evidence obtained from the scene of a crime. For example, much of the work of public analysts centred on the adulteration of substances, especially food.

2 K. D. Watson, *Forensic Medicine in Western Society: A History*, Abingdon: Routledge, 2011, p. 4.

3 Ibid., pp. 63–70.

4 Major studies of the history of scientific criminology often do not emphasize forensic science (e.g., see N. Davie, *Tracing the Criminal: The Rise of Scientific Criminology in Britain, 1860–1918*, Oxford: Bardwell Press, 2005.) This tends to mask potential influences of criminology on the development of forensic science.

5 A. Desrosières, *The Politics of Large Numbers: A History of Statistical Reasoning*, Cambridge, MA: Harvard University Press, 1998, p. 10.

6 See D. M. Vyleta, *Crime, Jews and News: Vienna 1895–1914*, New York and Oxford: Berghahn, 2007, 17–27; I. Burney, and N. Pemberton, 'Making space for criminalistics: Hans Gross and fin-de- siècle CSI', *Studies in History and Philosophy of Biological and Biomedical Sciences*, 2013, 44(1): 16–25.

7 Vyleta, *Crime, Jews and News*, p. 17.

8 H. Gross, *Handbuch für Untersuchungsrichter als System der Kriminalistik*, 2 volumes, Munich: J. Schweitzer Verlag, 1893. English language editions are considered later in this chapter.

9 P. Becker, 'The criminologists' gaze at the underworld: Toward an archaeology of criminal writing', in P. Becker and R. F. Wetzell (eds), *Criminals and Their Scientists: The History of Criminology in International Perspective*, Cambridge and New York: Cambridge University Press, 2006, 105–133.

10 J. W. Osterburg, and R. W. Ward, *Criminal Investigation: A Method for Reconstructing the Past*, Waltham, MA: Anderson, 2014, p. 18, 7th edition.

11 J. Adam, and J. C. Adam, *Criminal Investigation A Practical Handbook for Magistrates, Police Officers and Lawyers*, Madras: A. Krishnamachari, 1906; J. Adam, and J.C. Adam, *Criminal Investigation: A Practical Handbook for Magistrates, Police Officers and Lawyers*, London: The Specialist Press, 1907 (reprint of 1906 edition); J. C. Adam, *Criminal Investigation: A Practical Textbook for Magistrates, Police Officers and Lawyers*, London: Sweet and Maxwell, 1924. The first edition (1906) was reprinted in the UK in 1907, making it available to British readers. The first edition was also reprinted in the USA by The Lawyers' Co-operative Publishing Company, New York in 1907.

12 H. Gross, *Criminal Psychology: A Manual for Judges, Practitioners, and Students by Hans Gross*, 1911, London: Heinemann (translated from the fourth German edition by Horace M. Kallen).

13 T. Dixon, *Science and Religion: A Very Short Introduction*, Oxford: Oxford University Press, 2008.

14 M. J. Lynch, 'The power of oppression: Understanding the history of criminology as a science of oppression', *Critical Criminology*, 2000, 9 (1/2): 144–152, p. 147.

15 Ibid., p. 148.

16 Ibid., p. 149.

17 Ibid., p. 148.

18 D. Garland, 'The criminal and his science: A critical account of the formation of criminology at the end of the nineteenth century', *British Journal of Criminology*, 1985, 25 (2): 109–137, p. 109.

19 P. Jenkins, 'Varieties of Enlightenment criminology: Beccaria, Godwin, de Sade', *British Journal of Criminology*, 1984, 24 (2): 112–130, p. 112; C. Beccaria, *On Crimes and Punishments* (1764) (English translation 1767), 1767, available at https://archive.org/details/anessayoncrimes00beccgoog, accessed 24 January 2013.

20 J. Bentham, *A Fragment on Government*, Payne: London, 1776. Available at http://www.constitution.org/jb/frag_gov.htm, accessed 12 January 2015.

21 Jenkins, 'Varieties', p. 113.

22 Ibid., p. 128.

23 M. Foucault, *Discipline and Punish: The Birth of the Prison*, (trans. A. Sheridan), London: Allen Lane, 1977.

24 Ibid. p. 300.

25 Desrosières, *The Politics of Large Numbers*.

26 A. Bertillon, *L'identité des Récidivistes et la Loi de Relégation*, Paris: G. Masson, 1883.

27 C. Lombroso, *Criminal Man*. (Translated by M. Gibson and N. H. Rafter). Durham: Duke University Press, 2006. (original publication date 1876); F. Galton, 'Composite portraits', *Nature*, 1878, 18, May 23: 97–100.

28 C. Goring, *The English Convict: A Statistical Study*, London: HMSO, 1913.

29 Garland, 'The criminal and his science', p. 134.

30 Desrosières, *The Politics of Large Numbers*.

31 W. Morrison, *Criminology, Civilisation and the New World Order*, Abingdon and New York: Routledge Cavendish, 2006, p. 61.

32 Ibid., p. 62.

33 L. Daston and P. Galison, *Objectivity*, New York: Zone Books, 2007.

34 L. Daston and P. Galison, 'The image of objectivity', *Representations*, 1992, 40: 81–128.

35 Desrosières, *The Politics of Large Numbers*, p. 9.

36 Ibid., p. 10.
37 M. Lynch, S. A. Cole, R. McNally and K. Jordan, *Truth Machine: The Contentious History of DNA Fingerprinting*, Chicago, IL and London: University of Chicago Press, 2008.
38 Morrison, *Criminology*, p. 72.
39 Desrosières, *The Politics of Large Numbers*, p. 68.
40 T. Nagel, *A View from Nowhere*, Oxford: Oxford University Press, 1986.
41 Desrosières, *The Politics of Large Numbers*, p. 78.
42 Lombroso, *Criminal Man*.
43 J. Van Wyhe, *Phrenology and the Origins of Victorian Scientific Naturalism*, Aldershot and Burlington, VT: Ashgate, 2004.
44 Davie, *Tracing the Criminal*, p. 19.
45 Ibid.
46 M. J. Wiener, 'Murderers and "reasonable men": The "criminology" of the Victorian judiciary', in P. Becker and R. F. Wetzell (eds), *Criminals and Their Scientists: The History of Criminology in International Perspective*, Cambridge and New York: Cambridge University Press, 2006, 43–60.
47 Ibid., p. 45.
48 Becker, P. and Wetzell, R. F. (eds) *Criminals and Their Scientists: The History of Criminology in International Perspective*, Cambridge and New York: Cambridge University Press, 2006, p. 3.
49 S. Shapin, *The Scientific Revolution*, Chicago, IL and London: University of Chicago Press, 1996.
50 P. J. Bowler, *The Non-Darwinian Revolution: Reinterpreting a Historical Myth*, Baltimore, MD and London: The Johns Hopkins University Press, 1988.
51 Ibid. p. 5.
52 S.A. Cole, *Suspect Identities: A History of Fingerprinting and Criminal Identification*, Cambridge, MA: Harvard University Press, 2001.
53 F. Galton, *Hereditary Genius: An Inquiry into Its Laws and Consequences*. London: Macmillan, 1869.
54 M. Bulmer, *Francis Galton: Pioneer of Heredity and Biometry*, Baltimore, MD and London: The Johns Hopkins University Press, 2003.
55 Davie, *Tracing the Criminal*, pp. 96–110.
56 Ibid., p. 92.
57 D. Novak, *Realism, Photography and Nineteenth-Century Fiction*, Cambridge: Cambridge University Press, 2008.
58 Daston and Galison, 'The Image'.
59 Davie, *Tracing the Criminal*, p. 99.
60 Ibid., p. 100; J. R. Ryan, *Picturing Empire: Photography and the Visualization of the British Empire*, Chicago, IL: University of Chicago Press, 1997.
61 F. Galton *Generic Images*, London: William Clowes and Son, 1879, p. 5 quoted in H. Grimes, *The Late Victorian Gothic: Mental Science, the Uncanny, and Scenes of Writing*, Farnham and Burlington VT: Ashgate, 2011, p.50.
62 A. Sekula, 'The body and the archive', *October*, 1986, 39 (Winter): 3–64, p. 48.
63 Ibid.; F. Galton, *Inquiry into Human Faculty and Its Development*, London: Macmillan, 1883.
64 Sekula, 'The body', p. 50.
65 Ibid., p. 53.
66 G. Pavlich, 'The subjects of criminal identification', *Punishment & Society*, 2009, 11(2): 171–190, p. 171.

67 Ibid.
68 Ibid.
69 Cole, *Suspect Identities*, p. 3.
70 Pavlich, 'The subjects', pp. 179–180.
71 P. J. Hutchings, 'Modern forensics: Photography and other suspects', *Cardozo Studies in Law and Literature*, 1997, 9 (2): 229–243, p. 239; also see Pavlich, 'The subjects', pp. 176–177.
72 Lombroso, *Criminal Man*.
73 Morrison, *Criminology*, p. 137.
74 Ibid., p. 101. The *craniograph Anfossi* traces the cranium outline; if this was not available calipers akin to forceps used in childbirth, the *pelvimeter*, could be used to measure cranium width. See Lombroso, *Criminal Man*, pp. 239–240. The *volumetric glove* was a special *gutta percha* glove attached to an indicator on a scale, effectively amplifying the signal; this was designed to make the pulse more readily visible. See Ibid., p. 224.
75 Lombroso, *Criminal Man*.
76 Morrison, *Criminology*, pp. 105, 137.
77 N. H. Rafter, *Creating Born Criminals*, Urbana and Chicago: University of Illinois Press, 1997, p. 110.
78 Ibid., pp. 114–115.
79 H. Ellis, *The Criminal*, New York: Scribner and Welford, 1890.
80 Rafter, *Creating*, p. 120.
81 Ibid., p. 124.
82 Ibid., p. ix.
83 V. Brignall, 'The eugenics movement Britain wants to forget', *New Statesman*, 9 December 2010, available at http://www.newstatesman.com/society/2010/12/british-eugenics-disabled, accessed 23 February 2014.
84 Vyleta, *Crime, Jews and News*, p. 14.
85 Ibid.
86 D. Garland, 'British criminology before 1935', *British Journal of Criminology*, 1988, 28 (2): 1–17, p. 2.
87 Ibid.
88 Ibid.
89 Ibid., p. 5.
90 Goring, *The English Convict*.
91 Davie, *Tracing the Criminal*, p. 128.
92 J. Devon, *The Criminal and the Community*, London: John Lane/Bodley Head, 1912, p. 13, quoted in Davie, *Tracing the Criminal*, p. 155.
93 Pavlich, 'The subjects'.
94 Vyleta, *Crime, Jews and News*, p. 15.
95 Ibid., p. 17.
96 Burney and Pemberton, 'Making space'.
97 E.g., see C. Evans, *Crime Scene Investigation (Criminal Investigations)*, New York: Chelsea House, 2009.
98 B. Godfrey, *Crime in England 1880–1945: The Rough and the Criminal, the Policed and the Incarcerated*, Abingdon and New York: Routledge, 2014, pp. 136–139; P. Rawlings, *Policing: A Short History*, Abingdon and New York: Routledge, 2014 (first published 2002 by Willan Publishing), pp. 177–180.
99 Gross, *Handbuch*. Note that R. Grassberger, 'Pioneers in criminology XIII. Hans Gross (1847–1915)' *Journal of Criminal Law, Criminology, and Police Science*, 1956, 47(4): 397–405, gives the initial date of publication as 1883 (p. 399) then 1893 (p. 405).

For others describing crime scene management, see, e.g., A. Lucas, *Legal Chemistry and Scientific Criminal Investigation*, London: Edward Arnold, 1920; A. Lucas, *Forensic Chemistry*, London: Edward Arnold, 1921; W. Jago, *A Manual of Forensic Chemistry Dealing Especially with Chemical Evidence, Its Preparation and Adduction. Based upon a course of lectures delivered at University College, University of London*, London: Stevens and Haynes, 1909; R. A. Reiss, *Manuel de Police Scientifique (Technique): I. Vols et Homicides*, Lausanne: Payot, 1911.

100 Gross, *Criminal Psychology*.
101 Becker, 'The criminologists' gaze', p. 114.
102 Grassberger, 'Pioneers in criminology', p. 397.
103 Ibid., p. 398.
104 Becker, 'The criminologists' gaze', p. 113.
105 Ibid., p. 115.
106 Grassberger, 'Pioneers in criminology'.
107 Ibid. p. 399.
108 Ibid., p. 404.
109 Vyleta, *Crime, Jews and News*.
110 Becker, 'The criminologists' gaze', p. 115.
111 Ibid., p. 120; Vyleta, *Crime, Jews and News*.
112 Becker, 'The criminologists' gaze'.
113 Vyleta, *Crime, Jews and News*, p. 17. Vyleta states that the first edition was published in 1893.
114 Grassberger, 'Pioneers in criminology'. The final two editions were published after Gross's death. In what follows, I use *Handbuch* to delineate the German editions of Gross's work. To distinguish these from the English editions, I refer to the English translations of Gross's *Handbuch* as *Criminal Investigation* as this was the main title used in all five English editions.
115 Adam and Adam, *Criminal Investigation*, 1906.
116 Adam, *Criminal Investigation*, 1924.
117 R.L. Jackson, *Criminal Investigation A Practical Textbook for Magistrates, Police Officers and Lawyers*, London: Sweet and Maxwell, 1962, 5th edition.
118 Cole, *Suspect Identities*; C. Sengoopta, *Imprint of the Raj: How Fingerprinting Was Born in Colonial India*, Basingstoke and Oxford: Pan, 2003.
119 Adam and Adam, *Criminal Investigation*, 1906; Adam and Adam, *Criminal Investigation*, 1907; Adam, *Criminal Investigation*, 1924.
120 T. Muhlbacher, 'Hans Gross, the father of criminology, and Arthur Conan Doyle', in G. Heuer (ed.), *Sexual Revolutions: Psychoanalysis, History and the Father*, London and New York: Routledge, 2010, 55–64, p. 60.
121 D. Arnold, 'Crime and crime control in Madras, 1858–1947', in A. A. Young, *Crime and Criminality in British India*, Tucson: University of Arizona Press, 62–88, p. 77. Also see R. K. Arora and R. Goyal, *Indian Public Administration*, New Delhi: Wishwa Prakashan, 2011.
122 Adam and Adam, *Criminal Investigation*, 1907, p. vii.
123 Ibid.; H. Steinert, 'Fin de siècle criminology', *Theoretical Criminology*, 1997, 1(1): 111–129, p. 118.
124 Adam and Adam, *Criminal Investigation*, 1907, p 355.
125 Arnold, 'Crime and crime control'.
126 Adam and Adam, *Criminal Investigation*, 1906, p. 355.
127 Cole, *Suspect Identities*, p. 66.

128 Ibid., p. 67.
129 M. Radhakrishna, 'Colonial construction of a "criminal" tribe: Yerukulas of Madras Presidency', *Economic and Political Weekly*, 2000, 35 (28/29): 2553–2563.
130 M. Radhakrishna, *Dishonoured by History: 'Criminal Tribes' and British Colonial Policy*, Hyderabad: Orient Longman, 2001, p. 27.
131 Radhakrishna, 'Colonial construction', p. 2554.
132 Ibid., p. 2553.
133 Radhakrishna, *Dishonoured*, p. 82 quoting F. Booth-Tucker, *Muktifauj, or, Forty Years with the Salvation Army in India and Ceylon*: London and Edinburgh: Marshall Brothers, 1923, p. 226.
134 Radhakrishna, 'Colonial construction', p. 2554.
135 Cole, *Suspect Identities*, p. 69.
136 Adam and Adam, *Criminal Investigation*, 1907, pp. xxvi–xxvii.
137 See N. Kendal, *Criminal Investigation A Practical Textbook for Magistrates, Police Officers and Lawyers*, London: Sweet and Maxwell, 1934, 3rd edition; R. M. Howe, *Criminal Investigation A Practical Textbook for Magistrates, Police Officers and Lawyers*, London: Sweet and Maxwell, 1949, 4th edition; R. L Jackson, *Criminal Investigation: A Practical Textbook for Magistrates, Police Officers and Lawyers*, London: Sweet and Maxwell, 1962, 5th edition.
138 Adam and Adam, *Criminal Investigation*, 1907, p. xxv.
139 Ibid., p. xxvi.
140 Ibid., p xxii.
141 Ibid.
142 Ibid., p. xxiii.
143 Ibid. Public officials appeared in the *Gazette of India* and were classified as 'gazetted' if their rank was managerial or executive. Non-gazetted referred to ranks of middle-ranking professionals and below. Original italics.
144 Adam and Adam, *Criminal Investigation*, 1907, p. xxiii.
145 Walls, *Expert Witness*, p. 186.
146 R. B. Pilcher, *A List of Official Chemical Appointments in Great Britain and Ireland, in India and the Colonies*, London: Institute of Chemistry, 1906.
147 Becker, 'The criminologists' gaze', p. 117.
148 R. B. Pilcher, *A List of Official Chemical Appointments in Great Britain and Ireland, in India and the Colonies*, London: Institute of Chemistry, 1906.
149 Kendal, *Criminal Investigation.*
150 Ibid., p. xiii.
151 Howe, *Criminal Investigation*, p. v.
152 G. W. Wilton, 'Review of *Criminal Investigation* by Hans Gross, 4th edition, Edited by Robert Howe,' *Journal of Criminal Law and Criminology*, 1952, 43 (2): 281–282. In Scotland, the post of sheriff denotes a trial judge who deals with criminal and civil court cases.
153 Ibid., p. 281.
154 Ibid.
155 Jackson, *Criminal Investigation.*
156 Ibid., pp. v-vi.
157 Ibid., p. v. Note that Jackson used the term 'Investigator' rather than 'Investigating Officer'. However, the introduction to the second edition (Adam, *Criminal Investigation*, 1924) used 'Investigating Officer', and Jackson's edition reproduced the introduction to the second edition verbatim. Hence 'Investigating Officer' is Adam's terminology rather than Jackson's.

158 See Lucas, *Legal Chemistry;* Lucas, *Forensic Chemistry*, 1921.
159 Jackson, *Criminal Investigation*, p. vii.
160 Ibid. Original italics.
161 Ibid.
162 Lucas, *Forensic Chemistry*, 1921; C. A. Mitchell *Documents and Their Scientific Examination: With Especial Reference to the Chemistry Involved in Cases of Suspected Forgery, Investigation of Disputed Documents, Handwriting, etc.*, London: C. Griffin, 1922; J. L. Mnookin, 'Scripting expertise: The history of handwriting identification evidence and the judicial construction of reliability', *Virginia Law Review*, 2001, 87 (8): 1723–1845; A. S. Osborn, *Questioned Documents: A Study of Questioned Documents with an Outline of Methods by which the Facts May be Discovered and Shown*, Rochester, NY: The Lawyers' Co-operative Publishing Co., 1910.
163 Mnookin, 'Scripting expertise'.
164 Jackson, *Criminal Investigation*, p. viii.
165 Ibid., p. viii.
166 K. Simpson, *Taylor's Principles and Practice of Medical Jurisprudence*, London: J. & A. Churchill, 1965, 2 vols., 12th edition.
167 Wilton, 'Review of *Criminal Investigation*'.
168 Lucas, *Legal Chemistry*; Lucas, *Forensic Chemistry*, 1921; *Forensic Chemistry and Scientific Criminal Investigation*, London: Edward Arnold, 1931, 2nd edn., 1935 3rd edition, 1945 4th edition. Lucas claimed that his 1921 first edition of *Forensic Chemistry* was a substantial revision of *Legal Chemistry*.
169 M. Gilberg, 'Alfred Lucas: Egypt's Sherlock Holmes', *Journal of the American Institute for Conservation*, 1997, 36 (1): 31–48.
170 Ibid.
171 H. J. Walls, 'The Forensic Science Service in Great Britain: A short history', *Journal of the Forensic Science Society*, 1976, 16: 273–277, p. 275.
172 Jago, *A Manual of Forensic Chemistry*.
173 E.g., Lucas wrote for *Annales du Service des Antiquités de L'Égypte*.
174 Lucas, *Forensic Chemistry*, 1921.
175 See M. Mazévet, *Edmond Locard: Le Sherlock Holmes Français*, Lyon: Editions des Traboule, 2006.
176 R. A. Reiss, *Manuel*.
177 H. J. Walls, *Forensic Science*, Sweet and Maxwell: London, 1968, p. 193.
178 Keith Simpson's Foreword to N. Morland, *Papers from the Criminologist*, London: Wolf Publishing Ltd., 1971, pp. 11–13. Simpson praised the high professional quality of Morland's journal *The Criminologist*.
179 E.g., see N. Morland, 'Review: *Science Against Crime*', *Journal of the Forensic Science Society*, 1974, 14 (1): 77–78.
180 See N. Morland, *Fingerprints: An Introduction to Scientific Criminology*, London: Street & Massey, 1936; N. Morland, *The Conquest of Crime*, London: Cassell and Co., 1937; N. Morland, *An Outline of Scientific Criminology*, London: Cassell and Co., 1950.
181 See Morland, *An Outline of Scientific Criminology*, p.112.

3 Technoscience and the technologies of criminal identification

Introduction

Chapter 2 introduced scientific criminology and its influence on the British approach to recognizing criminals, particularly habitual criminals, acknowledging the multi-faceted question of criminal identification as one of the most important issues facing police authorities. This was one way in which science could be brought to bear on crime, but as a physiognomic and evolutionary approach to science and the criminal, it did not make use of the analytical capacities of the physical and biological sciences. In contrast to a criminalistic approach, the abstract attempts of scientific criminology to identify purported criminal predispositions not grounded in material evidence offered few practical instructions to scientific detection and policing for performing criminal identification.

This chapter reviews a number of factors contributing to the development of scientific policing in relation to the problem of criminal identification, focusing on technologies, the anthropometric approach and fingerprinting. The development of scientific policing relied heavily on enrolling appropriate underpinning technologies; hence the concept of 'technoscience' is particularly useful for understanding how science and technology could be seen as contributing to the development of detection and policing. A range of ideas imported from scientific criminology, even those of criminal appearance and physiognomy to some extent, was enfolded into scientific policing along with transport, media and communications technologies. The notorious Crippen case of 1910 is a paradigm example illustrating a broad technoscientific spectrum of concerns. These included criminal appearance and how this was used to construct guilt, the use of communications and media technologies, given that the use of wireless telegraphy in apprehending Crippen was widely hailed as the first example of its central role in an arrest, and the part the press played in portraying the use of these technologies in policing.[1] In addition, the way that the physical evidence was handled and the role of the medical experts, with Spilsbury as a central figure, means that the Crippen case runs the gamut of technoscientific policing.[2]

A consideration of the use of technologies in scientific policing must also involve the opposite face of the coin, namely the use of science and technology *by* the criminal. The fear that criminals would rapidly catch up with any newly introduced technology and use it for nefarious ends was a widespread and realistic concern. Returning to the chapter's central focus on the technologies of identification, the story of the 'Tichborne claimant', also involving questions of appearance, provides a paradigm example of how in the late Victorian period, an individual could maintain an apparently plausible claim to a fraudulent identity over a period of years to the consternation of the legal authorities and with the drain of the public purse in costly court cases.[3]

Given that physical appearance appeared to be so little help in identifying criminals and fraudsters, the anthropometric approach championed by Bertillon offered the promise of an exact way of measuring criminality, particularly in relation to the problem of identifying recidivists or habitual criminals. Bertillon is chiefly remembered for his anthropometric scheme, but his contribution was much wider, as he was a pioneer of forensic photography and crime scene management; his criminalistics achievements have been somewhat eclipsed by the story of *Bertillonage*.[4] Although forensic mythology would have us believe that *Bertillonage* was quickly overtaken by the superior and more accurate technology of fingerprint identification, anthropometry and fingerprinting were used side by side for a number of years.[5] Successful anthropometry and fingerprinting methods were completely dependent on information processing, notably an appropriate physical database and reliable filing and matching algorithms. The second half of this chapter briefly reviews the introduction and demise of *Bertillonage* and the swift ascendancy of fingerprinting for criminal identification. With the introduction of anthropometric and fingerprinting approaches, criminal identification had moved from physiognomy towards a biometric means of criminal identification. The biometric turn was accompanied by a turn towards the crime scene as the locus for scientific policing and detection and the question of incorporating scientific aids into policing, which is where Chapter 4 takes up the story.

'Scientific' and technoscience

The term 'scientific' was successfully enrolled into service in modern policing in the first half of the twentieth century, particularly in terms of the use of technologies. 'Scientific' as a metaphor for a modern and progressive approach is predicated on supporting technologies. The so-called 'linear model' of scientific and technological development which regards technology as applied science flowing in one direction from scientific developments has largely fallen out of favour in the field of science and technology studies.[6] Importantly, the abandonment of a linear model reinforces the point that unequivocal distinctions between science and technology cannot usefully be made. Instead, science and technology are better seen as inextricably woven into what can be termed 'technoscience'.[7]

In the present study, nowhere is the meshing of heterogeneous scientific, technological, political and organizational elements into technoscience more apparent than in the development of scientific policing. The term 'scientific' was often used to signal that something was to be seen as modern and forward looking, but that did not mean that 'scientific' could simply be equated with 'progressive'. What is understood as 'scientific' in relation to disciplines is fluid across disciplinary boundaries and across time and is not always positive.[8] As Bud argues, of the related term 'applied science', certain terms become key words or concepts in a culture; thus, it seems to have been with 'scientific' at the beginning of the twentieth century.[9] A search for definitions of 'scientific' in accounts of scientific policing and detection is likely to yield little; the meaning of 'scientific' tends to be taken for granted and was rarely explored. The fortunes of scientific thinking in wider culture were variable; one only has to consider the trajectory of public opinion on science though the nineteenth century to understand this point. In the Victorian era fears that science could be dangerous were overtaken by narratives of progress as the Victorian era witnessed the Great Exhibition of 1851 and the marvels of railway and telegraph technology, fuelled by the popular lectures and writings of scientists such as Huxley and Playfair.[10] But the status of science was more mixed in the twentieth century. The aftermath of the First World War saw considerable ambivalence over the role of science to the extent that it was no longer possible to expect the public to show unbridled enthusiasm for the marvels of science.[11] Despite this, the rhetoric surrounding the 'scientific aids' movement in policing displayed little doubt about the value of science as a positive force in crime detection, as the next chapter explores.

Arguments for scientific policing and scientific detection developed apace; a technique or technology was more likely to be seen as scientific if it was new. It is no accident that 'scientific management', a form of organization which was taken to be methodical, centralized, progressive and modern, was coming into vogue.[12] Although explicit references to scientific management are rare in scientific aids to crime detection and forensic science discourse, nevertheless there can be little doubt that the gleaming, modern scientific laboratories portrayed in photographs or in words in autobiographies or popular works on forensic science of the middle years of the twentieth century were paeans to scientific management. Science, or at long last, something that was actually called 'forensic science', could be organized 'scientifically'.[13] Alongside the laboratories, opportunities to reorganize and centralize detection and policing represented another example of scientific management in action.

The use of scientific, supposedly progressive, techniques in policing was highly dependent on the introduction of a number of technologies, including but by no means limited to, information and communications technologies. Not surprisingly, many of these technologies, as distinct and new, have completely sunk from sight in present times and a former technological marvel, which once might have been regarded as scientific on account of its technical novelty, may no longer be seen as such when the technology is used every day.[14] For instance, except when they do not work or when we feel that they are holding us up unduly, we rarely stop to reflect on the technological marvel that is the traffic light in the urban environments of the Western world. Yet traffic lights were a significant technological innovation in policing, particularly in cutting high road traffic accident rates in cities.[15] Similar innovations, once regarded as highly novel but now quotidian, include trains, motor cars and telephones. From where we sit, using motor vehicles or telephones in an organization's activities would hardly rank as technologically innovative compared with a specialized laboratory-based technique to aid the detection of crime. But that was not the case in the first part of the twentieth century where innovations in policing were as much dependent on transport and communications technologies such as cars, traffic lights and police boxes as they were on, say, the development of comparison microscopes and the use of ultra-violet light.[16] Sir John Maxwell CBE, former Chief Constable of Manchester, made this point in 1946 and his words nicely illustrate that 'scientific' meant the introduction of novel technologies just as much as it referred to what scientists do in laboratories:

> . . . great progress has been made in the application of science as an aid to police work. The introduction of the motor car . . . the use of wireless, the gradual development of police laboratories to help in crime investigation and the adoption of traffic signal lights are typical examples of the process of mechanization.[17]

Maxwell's paragraph is a manifesto for technoscience as the guiding light in scientific policing. Under the banner of 'scientific aids', a term which was often used in the 1930s and 1940s, the formative years of scientific policing and the forensic science laboratory network, he seamlessly wove together significant examples of scientific policing: cars, radios, traffic lights and police laboratories in 'the process of mechanization', thereby underscoring the significance of the whole spectrum of technological innovations in scientific policing.

Transport and communications technologies – criminals as scientists

The idea that a technological innovation to combat crime could, and would inevitably, be used by criminals was an important and recurring theme in contemporary discussions of the introduction of technology into policing. This reflected a perennial response to the introduction of all sorts of new technologies, which manifested itself as a 'moral panic' that new technologies would always be used for nefarious ends.[18] First the technology is hailed as a breakthrough, heralding a new age which is to be different from the past, and all sorts of ways in which the technology might be used are suggested; then there is panic about whether the technology could be used for negative ends. The reality often ends up somewhere in between. Technology is used, often very inventively, in ways that may not have been possible to predict, a point which has been made by science and technology studies authors arguing against technological determinism.[19] At the same time, all sorts of technological crime are imagined, and although criminals could develop strategies for using a new technology in crime, no one can necessarily predict how this might happen.

Nowhere was this kind of utopian/dystopian rhetoric in predictions of technology use more prevalent in recent times than in early discussions of computer and Internet use, but it was also common, more than a hundred years before, in reactions to the introduction of the telegraph, or, as Standage so aptly dubs it, *The Victorian Internet.*[20] We will consider discussions of policing, crime and communications technology later in relation to the Crippen murder case. Nevertheless, however tempting it is to think that the term 'technology' only refers to information and communication technologies, we should note that the wide range of technological interventions gradually introduced into English police forces in the 1920s and 1930s inspired similarly utopian/dystopian thinking early in their use, in particular, the concern that new technology would be adopted by criminals just as quickly as police authorities employed technologies for controlling crime; hence, it would always be a struggle for police forces to keep up with technologically savvy criminals.

Railways, the telegraph and then the telephone, followed by the automobile, offered unprecedented opportunities for crime and fraud. The idea of a 'professional criminal' able to exploit the opportunities afforded by the new technology steadily grew. As Inspector Bonfield from the Chicago police commented in relation to telephone fraud in 1888:

> It is a well-known fact that no other section of the population avail themselves more readily and speedily of the latest triumphs of science than the criminal class . . . The educated criminal skims the cream from every new invention, if he can make use of it.[21]

Then, as now, the educated criminal got his information about forensic technologies from the media; indeed, the reporting of the Crippen case delivered much useful intelligence about what to do to avoid detection by 'wireless'. Machado and Prainsack's contemporary empirical study of prisoners' attitudes to forensic technologies reveals that 'media portrayals were in some respects seen as practical sources of information and knowledge about criminal tactics'.[22] Beauregard and Bouchard describe this phenomenon as 'forensic awareness'.[23] Although they coin the term to describe contemporary behaviour, and explicitly describe it as a recent phenomenon, there is every reason to believe that earlier criminals would have displayed 'forensic awareness' of forensic and technological knowledge of their time. Certainly, the rhetoric employed by those fighting crime implied that part of the fight was against the scientific and technical knowledge of the criminal. As criminals were perceived to become more scientific, so, too, did means of detecting them. Science and technology could be a means towards professionalizing a career as a criminal, particularly for those who could pass themselves off as well to do in committing 'middle-class fraud'. Indeed this sentiment readily translated into 'it takes a scientist to catch a scientist' rhetoric which forensic chemist Alfred Lucas employed to promote the value of the new discipline and its practitioners:

> . . . in fact the criminal is becoming so scientific, not only in his work, but also in the means he adopts to escape detection, that a scientist is needed to cope with him, that is to say, a scientist must be set to catch a scientist.[24]

For instance, discussion on ways in which transport technologies were used by the police also had to include concerns about criminal use of such technologies and how these should be anticipated and thwarted. Rail travel inspired just such hopes and fears in terms of crime and crime control. The introduction of the railway was as much an innovation in crime as it was in policing, contributing to 'a radical shift in behaviour and mentality'.[25] Karstedt argues that in Germany, although the point also applies to the UK, the railway brought about a 'traffic revolution'.[26] The mobility of ordinary people was massively increased; they could remain relatively anonymous while travelling, and strangers were legitimately brought together in railway stations and train carriages.[27] The worry that a criminal could commit a crime then get away on a fast train was alleviated to some extent by the availability of the telegraph for sending a message to a police force further along the track. One of the first such uses was the apprehension of John Rawell in 1845. He murdered his mistress in Slough, put on a brown coat and took the train to London, where officers from Paddington Station, having been telegraphed his description, were able to make an arrest.[28]

In proposing the Banditry Bill to the UK Parliament in 1933, part of the intention was to thwart 'the criminal class [taking] advantage of such a wonderful and beneficent new invention as the motor car' as the Bill's sponsor, Hall-Caine, so eloquently put it in his description of the new crimes afforded by 'mechanically propelled vehicles'.[29] As the title of the Bill suggests, criminals availing themselves of mechanically propelled vehicles for their nefarious activities were not of the ordinary type; they were bandits. Banditry is a term which is usually associated with outlaws, people who were not ordinary criminals but were in the margins, beyond the law, redolent of the Wild West, and it was also a term used to describe groups of criminals in colonial India.[30] Although such an interpretation might be seen as reading too much meaning into the term, nevertheless calling car criminals 'bandits' does seem to encapsulate something of the fear associated with the criminal possibilities of new transport technologies to be in the margins and beyond the law.

A strong class and gender subtext was woven into the Bill, which was designed to give police stronger powers to stop and search motor vehicles. The House of Commons discussion warned of the perils of women being subject to robberies by drive-by criminals. However, they were not seen to be safe in their cars either, as driving after dark was regarded as especially perilous for women, lest they be apprehended by men at the roadside pretending to require assistance who would then turn out to be robbers. Commander Marsden noted that those living in the country had difficulty getting domestic servants, as the latter worried about having to walk 'a mile or two' to an omnibus or post office as they, too, feared they would be accosted by car-driving criminals.[31] Although the Bill failed, there were real concerns for the newly described motorized crime of 'smash-and-grab' to the extent that there was something of a moral panic in the UK over the possibility of gangs of criminals riding around in motor cars committing crimes and speeding away from the crime scene.[32] And, of course, the possibility of rising criminal mobility *was* increased and the term 'smash-and-grab' remains in the lexicon of crime. In response, UK police forces introduced police motorbikes and cars, including undercover police cars, or 'Q cars', and attended to police driver training, even enlisting famous racing driver Sir Malcolm Campbell to advise.[33]

Criminal appearance – the Crippen case

The Crippen murder case of 1910 acts as an exemplar, bringing together a number of themes from scientific criminology and the development of forensic science in terms of criminal appearance, new communications and media technologies, forensic analysis and the authority of the expert witness.

Physiognomy, the treatment of physical appearance, and how far it was understood as a marker of criminality was a central plank of late nineteenth-century scientific criminology. Although police forces issued descriptions of suspects, a science of criminal appearance was not central to the development of scientific policing, and, in a formal sense, physiognomy was a discredited theory. Nevertheless, in interesting ways the physical appearance of the criminal was translated and enfolded into scientific policing in the early twentieth century as the Crippen case illustrates, combining with other technological and scientific advances to confirm criminality. This suggests that ideas from scientific criminology could be imported in subtle and sometimes informal ways into policing even when the original concept had been rejected, particularly when strengthened by popular imagination and the press, then as now, on the lookout for a sensational story. Hence, although the concept of criminal physiognomy had fallen from favour by the early twentieth century, nevertheless both police and popular accounts drew upon the idea of a criminal look or appearance in constructing guilt.

The bones of the Crippen murder story are as follows. In 1910, Cora Crippen, a music hall artiste, mysteriously disappeared. Initial searches of the house she lived in with her husband, Dr Hawley Harvey Crippen, revealed nothing. Only when Crippen fled did the police search the house thoroughly to find body tissues buried in the basement of their London home. The tissues were only part of a body and not enough to confirm the gender, let alone the identity of the individual. Nevertheless, they were identified as those of Cora Crippen.[34] Because Crippen was not immediately arrested, he had been given the opportunity to make his escape with his lover, Ethel Le Neve, disguised as Mr and Master Robinson, respectively, bound for Canada on the *SS Montrose*. The captain of *SS Montrose*, suspecting that he had Dr Crippen and his lover disguised as a boy aboard ship, radioed the ship's owners and the police. This gave Scotland Yard detective, Walter Dew, who had not previously thought to arrest Crippen, the opportunity to board a faster ship which overtook Crippen's slower journey in order to apprehend the suspect couple as the *SS Montrose* docked near Quebec.[35]

The press and the public were eager to criminalize Crippen's appearance, and his physical description quickly moved from the mild oddity of the police description towards the bizarre. These descriptions were not direct attempts to revive Lombrosian criminology in the twentieth century (the descriptions derived from popular media rather than professional or academic sources), but they were, rather, convenient and acceptable ways of 'othering' Crippen and reinforcing the argument for his guilt in murdering his wife in this highly unusual yet to this day inconclusive case.[36] This was one of the ways in which Crippen's guilt was constructed, reinforced and achieved. His guilt was rendered all the more incontestable when the acknowledged role of new information and communications technologies were lauded as scientific crime-fighting tools. The Crippen case, then, demonstrated to dramatic effect the power of marrying older views of scientific criminological physiognomy, which perhaps had never completely gone away, with the new technologies rapidly being taken up by scientific detection.

The police issued a detailed description of the wanted man shortly after his disappearance, and if we consider this description, we have an example of the reason why Alphonse Bertillon, faced with the perennial problem of inexact descriptions of criminals and suspects in the Paris Prefecture of Police, tried valiantly to make descriptions standard through his *portrait parlé*, literally a 'spoken portrait' of precise descriptions and standardized language.[37]

Accurate description is still a problem for police forces around the world, and detailed verbal descriptions, including height, build, clothing, gait, mannerisms, etc., remain important. By any standards, Crippen was an unlikely looking criminal as this part of his police description attests:

> age fifty; . . ., 5ft. 3in.; . . . bald on top; long sandy moustache, rather straggly . . . eyes grey; flat on bridge of nose; false teeth; wears gold-rimmed spectacles . . . rather slovenly appearance; throws his feet out when walking; . . . . slight American accent; wears hat back of head; very plausible and quiet-spoken; speaks French; carries firearms; shows his teeth when talking.[38]

However, the police description hardly describes a Lombrosian born criminal, even if his appearance was somewhat unusual.

But oddities in Crippen's appearance then began to be magnified by the popular press to make it easier to cast him as a criminal. Such descriptions dwelt on his size, gait and, in particular, his eyes which were described as curiously bulging.[39] Indeed, the press had a field day over the unfortunate Dr Crippen's appearance, yet he can hardly have appeared stranger than his lover, Ethel Le Neve, when she dressed as a boy for the transatlantic voyage. Le Neve's police description portrays her as an attractive young woman without distinguishing features and nothing which could initially be regarded as out of the ordinary.[40] Crippen was found guilty and sentenced to be hanged; Le Neve was acquitted. Detective Walter Dew's memoir, although produced many years after the murder case, makes it clear that Le Neve was never regarded as a guilty party and makes no judgment over her appearance.[41] One report, concentrating on Crippen's quack doctor activities, coupled his poor clinical technique with his appearance. So physical appearance could be connected, not just directly with criminality, but also with wider moral failings. 'The carelessness, however, was only in keeping with the most damning evidence: Crippen's appearance. "His face and eyes told the story of a life miserably misspent." '[42] Although, as Early describes, a certain amount of public sympathy was directed towards Crippen and Le Neve as the trial progressed, during the chase and subsequent arrest, the press were all too keen to cast Crippen's appearance as degenerate, harking back to the evolutionary physiognomic descriptions of scientific criminology.[43]

For instance, other than describing his eyes as grey and his gold-rimmed spectacles, the initial police description said nothing else about Crippen's eyes. However, as time progressed one could be forgiven for imagining that his slightly protuberant eyes had grown to the extent of bulging out of his head. Protuberant eyes were not one of the features of Lombrosian criminal physiognomy which suggested instead that deep-set eyes were a sign of criminality.[44] It was as though a slightly unusual aspect of Crippen's appearance could be seized upon and magnified. A number of newspaper articles made much of his supposed bulging eyes, which could have been a symptom of a medical condition, and turned this into a sign of abnormality, indeed criminal abnormality. Note the Daily Mail's description: 'the glassy protruding eyes of some deep-sea creature'.[45] Ethel Le Neve's father declared that Crippen might be wearing goggles to disguise his bulging eyes and might be disguised as an old woman; another suggested his effete appearance would lend disguise as a woman likely.[46] Increasingly bizarre suggestions were made about how he might disguise himself. Characterizing Crippen as a man who could easily disguise himself as a woman was another way of making him seem unlike a 'normal' man, therefore abnormal and easier to cast as a criminal. As Early put it:

> Readers eager to read the criminal face would find the catalog of Crippen's physical peculiarities begging for fit with . . . degeneration theories . . . Lombroso's criminal types . . . With each theory or system suffused with 'irregularity,' the tie between a mutilator and a sexual degenerate was made.[47]

This was before it was known that Le Neve was dressed as his son. The *News of the World* added: 'It is said that his physical conformation lends support to the belief that he is a degenerate'.[48]

The popular treatment of Crippen's physical appearance stands as an example of how concepts developed as part of scientific criminology; indeed, the very significant concept of appearance signalling Lombroso's stigmata writ large could be twisted and imported into twentieth-century criminal justice and have the term 'degenerate' attached to make it clear what his physiognomy signified. His appearance was increasingly described as criminal and degenerate as the case progressed, and therefore in the eyes of the public and the police was a major influence in constructing Crippen's guilt. However, it was wireless telegraphy which made his arrest possible despite initial police errors.

Crippen and communications technology – the power of the press

The use of information and communications technologies in crime and policing provide a clear illustration of the hopes and fears for technology in criminal detection. The Crippen case illustrates one of the best-known early uses of wireless communication in crime detection, where Captain Kendall was able to alert the police, the press and, so it seems, everyone else, that he had the suspects aboard by the use of wireless telegraphy. It was not just the sleuthing activities of the captain which prompted telegraph messages.[49] In hot pursuit on the *SS Laurentia*, Inspector Dew was broadcasting wireless messages into the ether in the hope that Captain Kendall would receive them and reply.[50] These less-than-private messages let the world know that the chase was on and prompted a public photo opportunity as the ship arrived in Canada, captioned 'a Capture by Wireless'. . . 'new information technologies had been brought – stunningly – into the service of law and order'.[51] This referred to Captain Kendall's 'Marconigrams', basically a 'wireless telegram' – a message sent by means of Marconi's wonderful new wireless system to Scotland Yard. Utilizing an improvised 'photo fit' detection method, which was nothing short of ingenious, Kendall blocked out hair and moustache with cut-out cardboard on newspaper photos of Crippen and Le Neve.[52] This allowed him to match the two suspects to the father and son 'odd couple' on board his ship.

Of course, Kendall's radio exchanges with Scotland Yard, however, were far from private as they bounced from one ship relay to another. That the technology was badged as wireless telegraphy and the message was termed a 'Marconigram' illustrates the way in which a new technology in its early life may be named and described by analogy with an older technology to which it bears some resemblance or analogy so we can understand its affordances, even if it works in a substantially different way. For instance, witness the initial naming of DNA profiling as DNA fingerprinting. Of course, DNA profiling is scientifically nothing like fingerprinting, but the analogy lies in the ability to uniquely identify an individual made possible by the technology.

Wireless telegraphy may have been like the telegraph in its ability to deliver messages by electrical means, but the privacy implications were very different. It was, of course, possible to intercept a telegraph message, but the telegraph is not a broadcast technology. In this respect, wireless telegraphy was unlike the point-to-point technology of the relatively private cable telegraph. Instead, it was a broadcast mechanism which had all-too-obvious unfortunate privacy implications.[53] Newspapers eagerly printed diagrams with overlapping circles round ships in the Atlantic to explain to the public how the broadcast nature of this technology worked.[54] Indeed, the use of this new technology in the Crippen case brought into sharp relief the ways in which new information and communications technologies would alter distinctions between private and public forever. The press exploited the broadcast nature of the technology to full advantage and by the time Crippen's ship had docked and he and Le Neve had been arrested, it seems as though the whole world knew about the case, as it had been widely reported in the press. The fact that it had been turned into a public spectacle of international dimensions surely compromised the opportunity for a fair trial. The dramatic chase and arrest obscure the way in which the police case had been bungled in the first place, and obviously Dew's memoir gives no suggestion that he was wrong not to have arrested Crippen before he was able to abscond.[55] The new technology unbalanced the prior equilibrium between criminal and detective, giving law enforcement the upper hand.[56]

The case itself and how it was handled by the police raised public discussion of information privacy, which is still hugely pertinent to contemporary concerns about freedom of information and privacy in relation to crime and security. However, the press view was not necessarily that Scotland Yard *should* attempt to keep its communications private. *The New York Times* made it clear that Scotland Yard was too secretive, arguing that wireless would force criminal justice authorities to be more accountable and open to the public.[57] This would also have the effect of making the public more watchful of breaches of the law. In moral terms, wireless offered the promise of more egalitarian ways of handling and circulating information about crime, echoing the trope that new information and communication technologies will inevitably be more egalitarian.[58]

'Thus is the arm of the law lengthened in a dramatic fashion by one of the latest developments of science.'[59] Newspapers on both sides of the Atlantic reported daily on the events of the Crippen chase. Even if there was no new information to report, it was possible to let the public know that the news had been obtained 'By Marconi Transatlantic Wireless Telegraph to The New York Times', thus indicating that the newspaper was utilizing the most modern communication technology then available.[60]

The Crippen case was by no means the first to receive such press attention. From at least Victorian times, there was a huge appetite for crime reporting, particularly the story of investigation, arrest, court appearance and sentence – the 'unfolding drama' across a wide spectrum of titles aimed at different social groups. The new tabloids created at the turn of the century, *Daily Mail* and *Daily Express*, were well placed to capitalize on the public appetite for crime reporting.[61] In an age where literacy levels for reading were quite high and were arguably increased by newspaper reading, consumption of crime reporting was considerable. 'Dukes and dustmen all enthusiastically read about crime in a variety of newspapers aimed at such different markets.'[62] The broadsheets and tabloids, although reporting the Crippen trial in considerable detail, tended to cover it in different ways, with the former focusing on the trial process, whereas the latter concentrated on the personal details of the accused and the witnesses.[63] Although Crippen was an iconic case for crime journalism, space and focus limit my analysis of press reports of the Crippen case.[64] Rather, it is press reports of the novel technologies used and the press role in promoting these which are of particular interest. By the time of the Crippen case, wireless telegraphy was the only means of on-board communication for ships on the ocean, and it has been suggested that '[t]he Crippen saga did more to accelerate the acceptance of wireless as a practical tool than anything the Marconi company had previously attempted'.[65]

Although the Crippen case provides a dramatic illustration of the use of new communications technology in crime detection, it was a relatively rare example. Police were understandably reluctant to send a description of a suspect by telegraph so that foreign police could make an arrest upon the suspect's arrival. For all that new technologies such as telegraphy and wireless telegraphy could be used in scientific policing, as well as being used for criminal ends, the Crippen case also demonstrated that the press could, and did, avail themselves of the same new technologies that the police used, making for interesting developments in the sometimes difficult relationship between police and press, something with which we still contend today. The Crippen story was made all the more glamorous and exciting as, for the first time, newspaper readers on both sides of the Atlantic were able to devour the story almost as it happened.[66] The public were to have an early taste of the kind of (nearly) instantaneous news to which we have become accustomed. And a side effect of the unfolding of this exciting tale was to make Inspector Dew look far better a detective than he deserved.

But if Drew was a minor hero, the press made it clear that the real hero was the technology, and the way that Crippen was 'captured by wireless' loomed large in the press reports. The promise of the new technology of wireless telegraphy took centre stage: 'new technology had been brought – stunningly – in the service of law and order'.[67] The popular press's relationship with detectives was important, but so, too, was the press's role in valourizing and explaining the role of technoscience in detection to its readers. The title of the *Daily Mail's* article of July 25 was 'The long arm of "wireless"'. '[A]lmost certainly they reckoned without the long arm of wireless telegraphy. While Dr. Crippen was doubtless congratulating himself on the skill with which he had hoodwinked the police the very messages which spelt doom were vibrating through the ether.'[68] Two days later, noting that this was the first time wireless telegraphy had been used to identify a police suspect, the paper quoted a 'senior detective officer' claiming that wireless telegraphy would be an important part of future police work.[69]

The *Daily Mail's* editorial 'Science and the fugitive' on 30 July 1910 was almost breathless in its admiration for the power of science and technology in apprehending the criminal, and it is interesting to examine how this article enrolled new technology as a major tool and, at the same time, enfolded other new technologies in use by the newspaper.[70] This signalled the value of the press in criminal justice, demonstrating that a newspaper should be seen as modern and progressive for its use of new information and communications technologies, and it was these that gave the *Daily Mail* a privileged position in gaining (nearly) exclusive access to Captain Kendall's words.

Crippen knew nothing of the chase whilst still on board the *Montrose*; he was

> . . . unconscious of the purport of the etheric waves passing silently over their head . . . [T]he nightmare awakening to the possibilities of modern science will come. He will realise too late that he has been tracked by the invisible bloodhound of wireless telegraphy, and that his efforts to escape have been watched with ironic laughter by the world. His scent has been followed on the high seas.[71]

So not only was wireless telegraphy a long arm of the law, it was a bloodhound on the scent; two wonderful metaphors illustrating the beginning of wireless telegraphy's role in crime detection. Not only that – the ether was enrolled as a mysterious bearer of invisible or inaudible messages.

Alongside its praise for the role of science, the editorial made the *Daily Mail's* own claim to its major role in the capture of Crippen explicit. The newspaper continuously published photographs of Crippen with descriptions in its Continental edition, and it was this edition that Captain Kendall had acquired before the ship left from Antwerp.[72] The published photographs allowed him to identify the pair, and the *Daily Mail* was rewarded by his sending a long account to the newspaper, using wireless telegraphy, while still at sea.

In the same editorial, the *Daily Mail* made its case for enfolding modern newspaper technology with wireless technology in capturing the criminal, pointing out that it was no longer possible for criminals to disappear.

> Newspaper photographs reproduce the unerring presentment of his appearance. Photographic engraving and the rotary press bring accurate copies of his likeness to the notice of men and women. The method of telegraphing photographs, invented by Mr Thorne Baker, can be used to transmit his likeness overseas with almost the swiftness of thought. Wireless telegraphy enables the police to interrogate the captains of steamers afloat.[73]

In two sentences, four technologies were deftly woven together in the fight against crime. Photography was, of course, a well-established technology by 1910, but reproduction of photographs in newspapers only started to appear towards the end of the nineteenth century as they relied on the development of halftone photography.[74] The *Daily Mail* began printing halftone photographs in its newspapers in 1907.[75] Similarly, although the rotary press was by no means a new technology, having been available from the mid-nineteenth century, it allowed for the mass printing of newspapers cheaply and quickly and no doubt was an important aspect of the *Daily Mail* being able to produce its Continental edition. From 1905, the Continental edition had been produced in Paris with news relayed by telegraph and telephone.[76] The rotary press, coupled with the new ability of newspapers to produce photographs, meant that images of potential criminals could be widely circulated.

As the *Daily Mail* immodestly, but accurately, claimed:

> Now the modern newspaper enlists the whole world in the work of detection. Its services in the warfare with crime, when it is well and widely employed by the police, cannot be exaggerated. But for the Press and the photographs which it published, even so alert an observer as Captain Kendall might never have had his suspicions aroused. The modern criminal is caged in a prison, fashioned by science and invention, from which there is no exit.[77]

The one area where the role of technology may have been somewhat exaggerated is the optimistic allusion to Thorne Baker's transmission of photographs by telegraph. Thorne Baker, an early pioneer of phototelegraphy, was able to claim, in 1910, using his version of the 'telautograph' to transmit pictures:

> The telegraphed picture, at first looked upon as a marvel, now occasions little or no surprise, which means that it is sufficiently like an ordinary photograph to pass muster among the other pictures in the newspaper in which it appears.[78]

When it came to wireless telegraphy, Thorne Baker noted:

> The prospects opened up by a wireless method of transmission are, on the other hand, of an encouraging nature, as not only could long distances be covered at a high speed, but photographs of criminals could be telegraphed to ships fitted with a receiving apparatus, and sketches or plans could be transmitted between different sections of an army.[79]

The transmission device he used was what we would recognize as a type of facsimile machine where messages could be transmitted such that line drawings could be drawn out at the receiving end. It is difficult to see how an image in any way equal to the quality of a photographic image in a newspaper could be transmitted with existing technologies. The images in Thorne Baker's book are not particularly clear. He predicted that it was *probable* that halftone photographs would be able to be sent using his methods.[80] As he himself noted: 'Time alone will show to what extent wireless photo-telegraphy will be of value'.[81] However, the point was perhaps not so much that this was an example of technoscience enrolled in the capture of Crippen; rather, it was a further example of how inventive information and communications technologies could imaginably be used in the future for apprehending criminals.

Crippen – the scientific evidence

The Crippen case had its fair share of mistakes; clearly Crippen should not have been allowed to escape in the first place. But there were also 'forensic blunders' in that the crime scene had been contaminated by the use of carbolic powder, disinfectant had been used round the crime scene by the police and the remains were left in an unlocked room in Islington mortuary covered by white paper of which no samples were available to test for contaminants.[82] Such contamination of evidence would not, of course, have held up in court in later trials – the chain of custody had been broken in a number of places. The work of Gross and Reiss was only just becoming known in Britain, and it was more than two decades before 'scientific aids' would be applied to policing and detection with its emphasis on preserving scenes of crime.

The prosecution had assembled a stellar team of forensic experts from St Mary's Hospital: pathologists Augustus Pepper and Bernard Spilsbury and toxicologists William Willcox and Arthur Luff. Thanks to 'a masterful piece of forensic work' they identified the bodily parts in the cellar as those of Cora Crippen and showed that half a grain of hyoscine, a fatal dose of the poison, was present in the tissues.[83] Pepper was the first scientific expert called by the police, and he was the first to examine the remains in the cellar. Willcox, the Home Office Analyst, then received specimens for examination in his laboratory at St Mary's Hospital. Given that only a part of the body had been buried in the cellar, the issue was to obtain as much of the body as possible to facilitate an accurate estimate of the amount of any poison present.

It was already known that Crippen had purchased five grains of hyoscine. Willcox extracted alkaloid from the tissues and, using the traditional method of putting a few drops in a cat's eye, concluded that the subsequent dilation showed the presence of a mydriatic alkaloid, a class of drug of which hyoscine is a member. Further tests showed that this was hyoscine. Using an accepted method of extraction, he obtained 0.43 of a grain of hyoscine from all the available organs. Given that much of the body was missing, this allowed him to make a very conservative estimate of at least half a grain in the whole body, a fatal dose.[84] This was a highly novel piece of forensic toxicological work, as it was the first case where hyoscine was identified as the poison used to commit a murder.[85] Spilsbury confirmed the body tissues as those of Cora Crippen through identifying a piece of skin as containing an operation scar.[86] His commanding performance in such a high-profile case was a paradigm example of 'celebrity pathology', and it propelled him to fame.[87]

One of the interesting features of the Crippen case related to the good state of preservation of the body parts; this considerably aided the ability of Willcox to discover the hyoscine. Pepper gave evidence on this point at the trial. The body had most probably been buried with quicklime, which, he argued, was caustic. However, in damp environments quicklime becomes slaked lime, which definitely has a preservative effect.[88] It seems likely that Crippen believed that quicklime could decompose body parts – hence its presence round the body – but he was probably unaware of the preservative effects of slaked lime. However, it was Alfred Lucas who first undertook scientific experiments on the preservative effect of lime on dead bodies, given that he could find no record of such experiments ever having taken place when he wrote in 1921, and he believed that Pepper's statement of the effect of lime was not accurate.[89]

Using a set of dead pigeons buried in limes in varying states, he showed that contrary to the common belief, not only quicklime suddenly slaked, but also quicklime slaked only gradually (e.g., from the moisture naturally occurring in the ground) would desiccate the body and have a preservative effect.[90] In his experiment, after examining the pigeon bodies after six months' burial, both the pigeon buried in quicklime *and* the pigeon buried in quicklime which had been suddenly slaked by adding water, were desiccated and showed no putrefaction.

> These results bear out the statements already made, namely first, that lime is a preservative, and secondly that the act of slaking lime in contact with a dead body, whether slaking is brought about gradually or done all at once, does not destroy the body.[91]

Willcox alluded to the preservative effects of the action of quicklime in a 1924 lecture on the Crippen case; nevertheless, his reply to a query from a member of the audience, who had attended the lecture and who quoted Lucas's experiments to Willcox, indicated that the latter had either not read about Lucas's results or he did not wish to acknowledge their significance.[92] Apart from the importance of the action of lime in helping to form adipocere, a fatty substance which forms a preservative coating round the viscera, Willcox alluded to the antiseptic effect of quicklime as it is alkaline, rather than to the effects of desiccation as a preservative, which was the function emphasized by Lucas in his pigeon experiments.

In sum, the Crippen case was remarkable for the way in which a breath-taking array of technoscience was enrolled to apprehend him, construct his guilt, bring him to trial and condemn him. These included old ideas on criminal physiognomy which were used to identify Crippen as displaying degenerate criminality, wireless telegraphy to confirm that he was aboard ship and the promise that photographs could be sent via telegraph, media technology in terms of quality photographic reproductions in newspapers, the mass circulation of newspapers made possible by the rotary press, the publication of the *Daily Mail*'s Continental edition facilitated by telegraph and telephone (which meant that Captain Kendall saw the relevant edition before setting sail), not to mention the forensic analyses at the trial, and Lucas's later demonstration of the preservative effects of quicklime. Were we to concentrate solely on the trial and the detection of poison and identification of the scar – in other words, the more properly forensic medicine or forensic toxicology parts of the trial – we would overlook the role of all these other technologies, all essential elements in the way the case ran its course. Early in the twentieth century, the public, press, police and criminal justice authorities had been handed a paradigm example of the way that science and technology were surely crucial partners in the development of scientific policing and detection.

Bertillon, identification and scientific policing

Although the Crippen case may have offered a stunning example of scientific policing's ability to marshall the powerful forces of science and technology in the fight against crime, it was not a case which centred on the perennial headache for scientific policing, namely the question of unique identification of the criminal. All over Europe the impetus to establish unique identity came from modern, centralizing bureaucracies.[93] Then, as now, there was a considerable civil component to this aim – taxation, commerce, movement across national borders and within countries, entitlement to pensions and benefits and so on. But there were also clear criminal requirements. Apart from the problem of reliably associating a criminal to a scene of crime, there was also the question of fraud. Fraudulently claiming to be someone else for gain was a matter of concern and an ever-present danger, as a number of well-known inheritance fraud cases attested.[94] One of the most notorious British fraudulent inheritance cases in the mid-Victorian period was that of the 'Tichborne claimant'. This involved an individual who was completely unrelated to the heir to the Tichborne estate but who managed to maintain his deception over a period of years and through two court cases before finally receiving a custodial sentence. Although he may have become a figure of fun, his case nevertheless was a salutary demonstration that, even in the apparently most civilized nation of the world, in the high Victorian period with all the trappings of science, technology and state, it was *still* not possible to accurately identify an individual, and it was *still* possible for a determined individual to maintain a fraudulent identity, often for years, at great expense, time and trouble for all concerned.

The ending of transportation of habitual criminals in 1867 and the setting up of a Habitual Criminals Register in 1870, with accompanying legislation, which allowed for reoffenders to be placed under police supervision for seven years in addition to any sentence passed, marked a pressing need to reliably identify repeat offenders in the UK.[95] 'Whether or not the criminal was a biological, rather than a social, entity, the Victorian penal system had a new incentive to identify the deviant, and so differentiate him or her from the respectable.'[96]

A potential solution to the problem of unique identification was to be developed in Paris. The development of a stable, orderly society in France's Third Republic was the backcloth to the rise of Alphonse Bertillon's anthropometric system.[97] To achieve an orderly society, it was necessary to identify and deal with criminals appropriately, particularly those who were habitual criminals. Recidivism was perceived to be as much of a problem in late nineteenth-century France as it was in the UK. In France, legislation allowed for the rehabilitation of first offenders, whereas repeat offenders were subject to particularly harsh punishments.[98] The problem, of course, was how to reliably identify the individual. If a criminal changed his or her name when caught and could not be linked to an earlier charge, then criminals were simply reinventing themselves as first offenders. Although photography had been introduced by the Paris police in 1872, it was used chaotically.

> [S]uspects were photographed, but very unsystematically – from varying angles and with varying degrees of expertise – and the prints accumulated in disorderly piles. They were catalogued by name alone, which meant that they were useless for identifying any malefactor who simply changed his name.[99]

Small wonder that Bertillon's systematic, anthropometric approach to identification was seen as a vast improvement to current police practices in Paris.

Bertillon is remembered (when he is remembered) for his anthropometric system of recording detailed measurements of the body which was adopted all over the world, including parts of South America, USA, UK, Russia, India, Germany and Switzerland. However, his contribution ranged even more widely than his anthropometric reputation would suggest, and this is why he deserves a central place in the history of scientific policing.[100] He was as much concerned with the accurate recording of crime scenes as with the indexing and classifying of criminal records; the latter was essentially an information technology and database management problem. We can see that Bertillon was a grand master of technoscience in the service of criminal justice, bringing together information technological management with scientific measurement.

His contribution can be seen in terms of his anthropometric system, the development of precise photography of the face and profile (he was probably the first to use photography of the profile in criminal identification), a precise coded way of describing the individual criminal's body, or, as he termed it, the *portrait parlé*, indexing and retrieval of criminal records and crime scene photography.[101] Although he had a notoriously difficult temperament, his contacts, colleagues and influences throughout the nascent forensic science disciplines were numerous. A number of Continental forensic scientists worked with him at early stages of their careers. These included Edmond Locard (1877–1966), Director of the famous Lyon forensic laboratory, and Archibald Reiss, who worked with Bertillon in Paris and, as Professor of Forensic Science at the University of Lausanne, went on to found the *Institut de Police Scientifique* there in 1909.[102] On the negative side, Bertillon's role in erroneous handwriting identification in the Dreyfus affair, always a difficult area for forensic science and a skill for which he had no expertise, was a considerable blow to his reputation.[103]

Bertillon came from a family of statisticians and demographers. His father was closely associated with Adolphe Quetelet. 'The Bertillon children grew up in a home filled with measuring tools, such as calipers and gauges, and a fervent belief in the importance of Quelelet's "social physics." '[104] They were therefore great believers in measuring people and applying statistical techniques to those measurements. But Bertillon was not a scientist and he developed no specifically scientific techniques. He was a measurer, sorter and classifier *par excellence*, and he arrived to sort out criminal identification technologies in what was surely a prime example of *cometh the hour, cometh the man*. On his appointment as a clerk at the Paris Prefecture in 1879, he quickly saw that the unruly way in which the Paris police held their photographic records militated against reliable criminal identification, and although he had to expend much effort convincing his superiors, he set to work to rectify the situation.[105]

Modern science and technology studies reject the view that technological successes are built on superior design with linear progress to the current day, casting aside technological failures on the way. Instead, we are exhorted to explain scientific and technological success in the same way that we explain scientific and technological failure; in other words, our explanations as to causes should be symmetric (i.e., the same kinds of explanations) with regard to truth and falsity, success and failure.[106] If we take a symmetric approach towards understanding Bertillon's work, we can avoid the temptation to regard *Bertillonage*, as his system was called, as an inevitably unworkable system which was bound to be overtaken by easier, cheaper and more accurate fingerprinting techniques. Instead, we need to explain the success of the system which was used alongside fingerprinting for a number of years, and we need to understand the place of *Bertillonage* in the wider ambit of information management and crime scene recording.[107]

Bertillon's system included measurements of eleven parts of the body, standardized photographs of the face, from the front and in profile, and the *portrait parlé*, which offered descriptions in an abbreviated code of visual aspects of appearance and a description of peculiar marks. The measurements had to be made in a defined way so as to optimize the chance of obtaining the same reading in future measurements. For instance, foot length was measured by having the subject lean forwards while standing so as to obtain maximum extension of the foot. Of course, this is not the natural way that foot length is measured. For instance, think of having one's feet measured for a pair of shoes.[108] Although the poses the subject had to adopt for *Bertillonage* were made for good reason, the unnatural way some measurements were taken no doubt contributed to the system's reputation for difficulty. The advantage of the *portrait parlé* was that, in addition to providing a standardized way of providing a description, the codes were abbreviated to allow for the possibility of telegraphing descriptions to other police forces. The final part of the system included two photographs, face on and in profile, taken in standard pose and lighting conditions.[109]

All the elements of the Bertillon system had to be carried out in a particular and precise way. This indicated one of its difficulties. Training by apprenticeship to the master would have transmitted the skills, including the *tacit knowledge* required to successfully replicate the procedure. But most people learned from manuals, which were far from simple, and this contributed to the difficulties of using the technique.[110] By way of comparison, in a very different scientific area and time, nonetheless with some similar features with regard to successful apprenticeship, Collins's classic study of laser building demonstrates apprenticeship to the master laser builders was vital for success in building a transversely excited atmospheric (TEA) laser.[111] Detailed scientific papers and scientific expertise were not enough to guarantee that physicists could build a successfully working laser. The process of apprenticeship transferred the tacit knowledge necessary for building a working instrument. Clearly, *Bertillonage* is not of the same order of complexity as laser building; nevertheless, the point about apprenticeship for successful skills transfer is one which is widely applicable in technoscientific work and must surely apply to the skills required in *Bertillonage*. One had to serve an apprenticeship to absorb the tacit skills necessary to make the system work.

Whatever the problems with implementing the full complexities of the Bertillon system, his approach captured the pressing need to systematize and code criminal records. In at least three important respects the Bertillon system involved sophisticated information and communications technology mechanisms that have been inscribed into the development of computing and information technologies in the twentieth century and are still highly relevant today.[112] Without judging the success of such an approach, the standardizing of a language for physical description is mirrored in twentieth-century computing's attempts to develop standardized languages for capturing user queries and requirements, notably 'structured English'.[113] Modern structured computing languages are designed to reduce ambiguity and to be computationally efficient. Bertillon's structured language attempted to do the same, for the same reason, although in Bertillon's case computational efficiency did not refer to the efficiency of an electronic machine; instead, it meant efficient in terms of the work of the human operator searching through physical files in a room. He understood that natural language is vague. He also understood that coding structured language descriptions would help efficient searches for records. Such codes also increase efficiency in the transmission of information by telegraph, in other words information compression. Indeed, coding information for telegraphic transmission was seen as extremely important for commercial and criminological reasons across a wide range of applications.[114] It allowed for precise messages to be constructed, thus reducing ambiguity; it afforded some measure of privacy (as one had to know the codes to translate the message); and, given that telegraph messages were usually charged on a per-word basis, it was a cheaper means of sending a message, as one word or abbreviation could stand for an agreed phrase. So Bertillon's coded descriptions were ideal for this purpose.

The final information technology aspect of the Bertillon system was the formation of a large, physically and conceptually, indexed searchable database. Once again, this was a precursor of modern database technology. He recognized that the right kind of classification was more important than homing in on individual markers.[115] Having collected an individual's information and written it on a card, he classified head length according to three categories – small, medium or large – sub-classified the cards successively according to head breadth, middle finger length and so on. Each set of sub-classified cards had its own drawer in the file. The idea was to end up with similar numbers of cards in each drawer. It was here that Bertillon's knowledge of population statistics came into its own. Appropriate definitions of small, medium and large were vital to achieve similar numbers in each category. He based these definitions on statistics from Parisian prisoners, as he knew how to divide up the populations under the normal distribution curve appropriately. The characteristic normal distribution for population statistics could be cut into three so that a narrow middle range was defined as the 'middle' and longer sparsely populated 'tails' of the curve could be defined as small and large, respectively. A new card could be matched against the room full of file cards that constituted Bertillon's database. If a potential match was found, the operator could then check the details for physical description and special marks.[116] Bertillon had constructed an efficient database management system to the extent that his system can be regarded not only as part of the story of science history and crime history, but also as an important plank in the edifice of information history.

Bertillon's system enjoyed considerable success in identifying repeating offenders from its introduction in the mid-1880s, thereby strengthening the belief that recidivists were indeed a significant problem and reinforcing the underlying tautology inherent in Galton's composite pictures. His system supported a belief that habitual criminals existed, which was what his system was created to find.[117] Indeed, it can be argued that recidivism was 'inscribed' in or designed into *Bertillonage*.[118]

Although Bertillon is chiefly remembered for his identification technologies, as Kaluszynski reminds us, his contribution to *criminalistique* was considerable through his research into methods for finding trace evidence at crime scenes and his developments in forensic photography.[119] The precision of his human photography was also applied to crime scenes in the form of *metric photography* with measuring scales so that there was a permanent precise record of the crime scene and of the location, placement and size of objects in a crime scene.[120] This was a vast improvement over sketches and descriptions. Once again, Bertillon's skill in measurement was clear. The great value of such photographs was that they allowed the size of a given object to be directly read off the photo with its accompanying scale. Bertillon's camera was fitted with an objective lens of exact and known focal length and with an exactly calibrated method for focusing. The camera was of precise known dimensions with ridges for photographic plates (30 cm × 30 cm), each 5 cm apart. When a photograph was taken, the camera was placed exactly horizontally at a distance of 50, 100 or 150 cm above the object or scene to be photographed. The dimensions of the apparatus, known focal length and distance from the object allowed the calculation of all measurements from the photograph. To simplify the procedure, Bertillon made special mounts printed with appropriate scales depending on the various parameters of distance, focal length and so on.[121]

Bertillon's metric photography was nothing if not ingenious, demonstrating once more his technical skill in exact measurement chanelled into a forensic technique, but one which, unfortunately, as with *Bertillonage*, demanded a considerable amount of knowledge and skill from its operators. As this outline implies, it was hardly likely to have been seen as a simple approach to photography. As a next-generation forensic scientist and admirer of Bertillon's photographic prowess, the Swedish forensic scientist Harry Söderman somewhat tellingly noted that the metric photograph was important but had never had its full value understood, as it was more difficult to take than an ordinary photograph. Although German, Austrian and Dutch scientists had tried to simplify the method:

> . . . it has fallen into disuse, and even in Bertillon's old laboratory in Paris it is probably used only in rare cases. This is deplorable, because it is often very important to be able to measure accurately the distance between different objects in a room as it appeared when the crime was committed.[122]

As a final, but important, coda to Bertillon's career, his anthropometric method had a resurgence in its application for recording the movements of travelling people in France. In this respect, the technique echoes Gross's interests in the control of gypsies and mirrors the development of fingerprinting and the colonial translation of Gross to control criminalized ethnic groups in India. Undeniably, forensic technologies were heavily implicated in the 'othering' of ethnic groups, particularly itinerant peoples. As in other European countries, so-called gypsies and nomads were treated with considerable suspicion in nineteenth-century France and, rather than advocating exclusion, public sentiment gradually turned towards detention and forcible integration in the early twentieth century.[123] An enquiry into the 1895 census of gypsies and nomads in France was tasked with investigating ways in which travelling people could be supervised so that detection of crime could be facilitated.[124] The introduction of an identity card gained widespread support, and this was formally settled by the passing of the Nomad Law in 1912 which included the introduction of an 'an anthropometric nomad passbook' or *carnet anthropometrique des nomads*.[125] This was a collective identity pass for families with pages for visas from areas where a family might temporarily stop, pages for anthropometric measurements including two photographs, face on and in profile, and fingerprints 'an instrument of intensive control that could justify an endless process of administrative harassment', reinforcing the 'criminal otherness' of gypsy groups.[126] So the fears and anxieties about itinerant populations in Europe and beyond were amplified and, at the same time, mitigated by the use of forensic technologies.

'Little worlds': the rise of fingerprinting[127]

The huge amount of interest in measuring and controlling the human body and, in particular, the racialized and criminal body, fuelled the interest in finding ways to incorporate scientific ideas into policing at the end of the nineteenth century and beginning of the twentieth.[128] Indeed, in Europe and North America particular links were made between the criminal body and different races. Nowhere is that story played out more clearly than in the history of fingerprinting.

The story of the rise of fingerprinting as a criminal identification technology has been told many times, and there are some excellent recent histories.[129] Hence, the story of the development of fingerprinting as a forensic technology is recounted here only in outline with the aim of completing the picture of technoscience in the service of forensic identification at the *fin de siècle*. Fingerprinting, once it was established as a practical forensic technology with appropriate record-keeping processes, gradually overtook *Bertillonage* as a cheap, apparently easy method which would permit unique identification of individuals, and which quickly came to be seen as an infallible method, thus going a long way to assuage the fears and concerns over criminal identification which the previous two chapters have emphasized.

Of course, people must always have wondered about the significance of the distinctive ridge patterns on the tips of their fingers, and there is evidence that they were used as a proxy for signatures from ancient times; hence it would be an exaggeration to say that they were 'discovered'.[130] Fingerprinting as a central method of reliable identification was born in India under British colonial rule. After the Indian Rebellion (or Sepoy Mutiny) of 1857 where much blood was shed on both sides, the British government took over the rule of India from the East India Company. As Sengoopta argues, race became even more of an issue for the British in India as they focused on charting Indian racial differences based on the caste system.[131] 'This apparently more "scientific" approach was stimulated less by the demands of science than by the 1857 Sepoy Mutiny.'[132]

The problem was how to reliably identify a unique individual for civil uses such as pensions and benefits, let alone criminal applications. This was a particular administrative issue in an environment where people were not literate and the need was compounded by the belief commonly held by the British colonial authorities that the indigenous people would naturally tend to be untruthful.[133] All these elements gave an added spur towards establishing more reliable means for accurately identifying the individual in colonial India. Recidivism may have loomed large in the minds of European police officials, yet their problem was nothing like the scale of the issue of identifying individuals entering into contracts or proving their identities for drawing pensions, a problem which could only be solved by administrators or anthropometric measurements. 'This was why systematic identification of fingerprints evolved in colonial Bengal rather than at the heart of London or Paris.'[134]

The story of systematic fingerprinting for civil identification in India started with William James Herschel (1833–1917).[135] Herschel was a colonial administrator and member of the famous dynasty of astronomers. Rather than following his illustrious forebears into astronomy, he chose a career in colonial administration, first in the East India Company and then in the Indian Civil Service.[136] The British administration was marked by its considerable distrust of the local people. Repudiation of contracts, once signed, was not uncommon; hence, in 1858, Herschel decided to ask a local contactor to 'sign' a contract with a handprint. His aim was to frighten the contractor out of the temptation to repudiate his signature later rather than a specific expectation that the handprint would be used for identification. From then onwards Herschel's fascination with fingerprints increased; he began collecting examples of fingerprints and studying their characteristics. He championed the cause of fingerprinting for contractual agreements during disturbances over the indigo industry, in which the plantation owners were implicated in many oppressive practices against the local growers.[137] Continuing to witness the huge number of legal cases on repudiation of contracts only served to strengthen his resolve that fingerprints were the answer. His first trials were with signatures on deeds and with pensioners. There was widespread suspicion that impersonators continued to collect pensions for years after the original beneficiary had died.[138] These trials were successful in that they resulted in a significant decrease in impersonation and repudiation. He also contacted the Inspector of Jails to ask him to use fingerprinting so that prisoners' identities could be confirmed. In spite of these successes, Herschel struggled with the Indian governing authorities who showed little enthusiasm despite his achievements and so his fingerprinting work had little lasting impact.[139]

An important aspect of Herschel's work was his huge collection of fingerprints which he had acquired over a span of many years. Indeed, he experimented on himself in this regard, showing that his fingerprints were unchanged between 1859 and 1916, as Figure 3.1 shows. He could not prove it, but he had much evidence to support the claim that fingerprints were unique to the individual and that they remained unchanged through life. This was exceedingly important evidence. Systems built on fingerprinting could only be reliable if these assumptions were accepted as reasonable.

The next major development in the fingerprinting story centres on the work of Henry Faulds (1843–1930), a medically trained Scottish missionary and keen

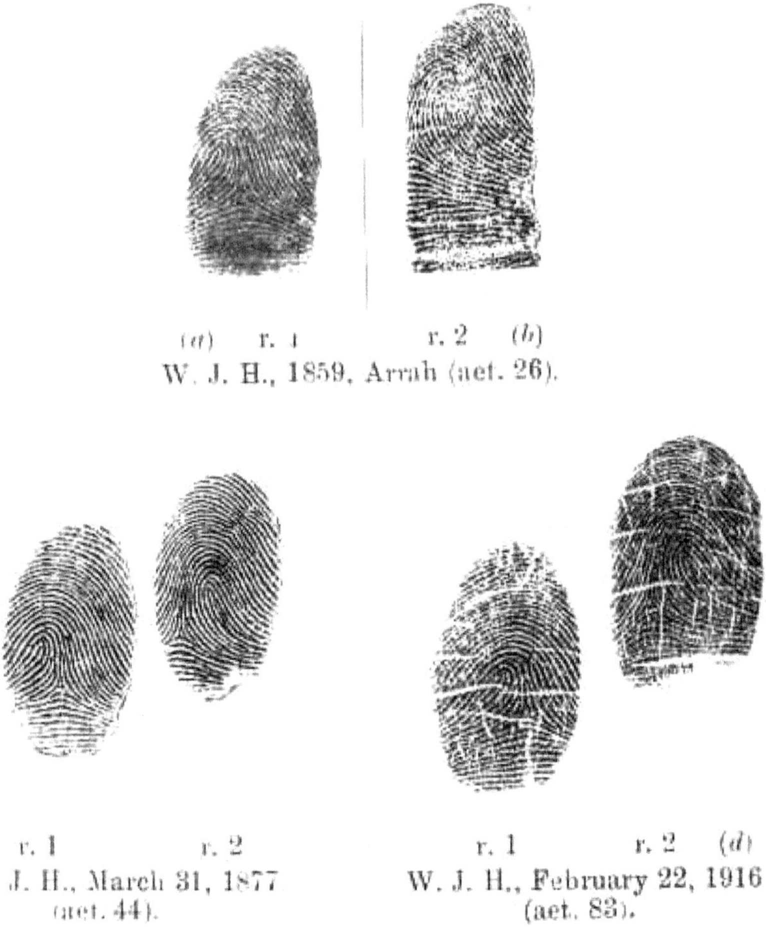

Figure 3.1 Herschel's fingerprints from 1859, 1877 and 1916.[140]

Darwinist who was working in Japan.[141] Having been prompted by his observation of fingerprints on ancient pots and hoping that fingerprint ridges would provide clues to markers of race, he began collecting fingerprints – even sending requests round the world for samples.[142] In 1880 he wrote to Charles Darwin, by then elderly and ill, to explain his discovery. In the letter he indicated the possibility of using fingerprints for criminal identification. Darwin forwarded the letter to his cousin Francis Galton. Meanwhile, Faulds wrote a letter to *Nature*, the foremost British scientific journal of the day, clearly signalling their potential in criminal identification.[143] This had the unfortunate but inevitable effect of triggering a priority dispute with Herschel.

Faulds may have had a better understanding of the importance of fingerprinting for criminal identification, but Herschel had copious evidence of the immutability of fingerprints.[144] Faulds failed to interest Scotland Yard and was virtually ignored by the establishment. He was an obscure Scottish doctor, far beyond the orbits of the scientific elite, and this must have contributed to his inability to convince the authorities to take fingerprinting seriously.[145] This was compounded by the fact that Galton overlooked his contribution and, of course, Galton had considerable power to influence the British establishment. Chapter 2 considered Galton's interest in the question of how far intelligence and ability was inherited, in particular, hereditary genius, to the extent that he published a book on the subject, *Hereditary Genius* (1869).[146] He was considerably influenced by position in society; hence, it was unsurprising that he enthusiastically corresponded with Sir William Herschel, also of appropriate scientific and gentlemanly lineage, whilst failing to acknowledge Faulds's contribution. The publication of his book on fingerprinting and his own personal prestige were vital in publicizing the possibilities inherent in fingerprinting.[147]

The earliest working fingerprint system for criminal identification is credited to Juan Vucetich in Argentina in 1891.[148] Scotland Yard adopted the system developed by Sir Edward Henry (1850–1931) in 1901.[149] Henry was the final significant character in the British fingerprinting story. He joined the Indian Civil Service in Bengal in 1873, rising through the ranks to become Inspector General of the Bengal Police in 1891.[150] Henry introduced the Bengal Police to anthropometry; anthropometric measurements were already used in India by anthropologists to measure differences between castes, but his interest lay in applying the system to the so-called 'criminal tribes' which were causing so much concern to the British administration.[151] Henry introduced a modified Bertillon system in 1893, with fewer measurements and including the left thumbprint – he had read Galton's monograph on fingerprinting by then.[152] Although he consulted and visited Galton and achieved reasonable success with his modified Bertillon system, he gradually came to believe that fingerprinting would be a better system if an appropriate classificatory scheme could be found. In 1897 he announced his fingerprint classification system.[153] This effectively sounded the death knell for *Bertillonage* in British India, and the system quickly spread. Henry was considerably aided by his two assistants, Azizul Haque and Hem Chandra Bose.[154] Haque, in particular, contributed greatly to working out the system to the extent where he is likely to have been its originator.[155]

In an important sense fingerprinting technology had similar information technology requirements to Bertillon's system, as both required a version of a 'one to many' search, i.e., the data from one individual searched against the data from many individuals for a match (the kind of relationship which is typically taught in elementary computer database courses). The database of cards where fingerprint images were stored, therefore, had to be organized in such a way as to categorize the cards accurately and efficiently so that the operator could focus on a group of cards where the match was likely to lie, thus cutting down very significantly on a lengthy 'brute force' search. In Bertillon's anthropometric system, his knowledge of population statistics made it possible for him to divide up his cards into reasonably equal sub-categories for efficient searching. For fingerprint searches, the challenge was to find a way of using the ridge patterns on the ten fingertips – loops, whorls, arches and composite patterns – so as to classify a set of prints accurately and in such a way that there were manageable numbers of records in each category. Because *Bertillonage* involved continuous numerical scales, categories of small, medium and large could be chosen to comply with known population statistics; this was not conceptually difficult. Fingerprint technology involves classification of discrete categories where the population's prints do not necessarily distribute themselves evenly over the different potential classifications, and this is where the challenge lies. The USA's first fingerprint classification system was inaugurated in New York in 1902, but in 1924, the FBI created a criminal file based on the Henry system. 'Over the next 47 years, the FBI manually arranged over 200 million fingerprint records into files using the Henry system of classification.'[156]

The Henry system involves an ingenious sub-classification of fingerprints over 1024 subcategories.[157] Of course, the sub-categories vary greatly in size from a single folder of cards to several drawers of cards. Large sub-categories could be sub-classified further. The system is far from infallible and depends on the accuracy of initial classification. An error in classification would be compounded through sub-classification. The trade-off is between reliability and speed and ease of searching for a match. Wayman et al. suggest that manual systems based on this classification achieved around 75 per cent reliability, and rigorous training of operators helped minimize indexing errors.[158] A match was identified by manually comparing the fingerprints with a magnifying lens to the search candidates identified by the system.[159] If a matching card existed, it was likely to be found, but not if the original prints had been misclassified or misfiled. So it was far from being an infallible system. It should also be remembered that the Henry system was similar to Bertillon's anthropometric system in that it relied on matching data from a suspect to data in the filestore to establish whether the suspect was already known to the authorities. Hence, these were both techniques for identifying recidivists. The Henry system relied on being able to obtain a complete set of 'ten prints' from a suspect, so it was not an approach which could be used to link a suspect to a mark left at a crime scene. The latter process relied on the ability to match a latent print or prints found at a crime scene to the prints taken from an individual suspect.

Early computer systems for fingerprint identification, introduced from the 1960s onwards, were based on the Henry system. These could overcome searching errors but not initial classification errors.[160] The introduction of computers takes us well beyond the scope of the present work; however, this serves to illustrate two things, namely just how long the system, developed through the efforts of a British colonial police official, was to survive, even to the extent of supplying the basic approach to automation and, second, that the classification continued to be dependent on the skill of the human fingerprint indexers.

Conclusion – technoscience and the dawn of scientific detection

The development of scientific approaches towards criminal identification was heavily dependent on supporting technologies, or 'technosciences'. Information, communications and transport technologies all held promise for enrolment into the armoury of scientific detection of crime; unfortunately, all of these held possibilities for criminal use as well. The potential for criminals to use new technologies, such as the motor car and telephone, only served to fuel the concern that the police needed to keep up with criminals who were becoming more 'scientific'.

Although it was, in several respects, inconclusive and ambiguities remain to this day, the remarkable Crippen case demonstrated to the police, criminal justice authorities and public that a wide range of technoscientific marvels, from the virtual 'bloodhound' that was wireless telegraphy, to a range of media technologies and forensic toxicology and microscopy were available to bring the criminal to justice. At the same time, police bungling over the initial stages of the investigation only served to underline the deficiencies of the police in scientific detection and their lack of ability to organize their operations to take command of the new technologies which, it seemed, were rapidly becoming available in the fight against crime.

Technologies for criminal and, indeed, civil and colonial identification were the other side of the technoscientific coin, with Bertillon's anthropometric system and fingerprinting as the main candidates. Both systems were heavily reliant on efficient and appropriate database management systems and search algorithms – in other words information technologies. Criminal identification was as much an information and communications technology problem as it was a forensic problem. In common with Gross's system of *criminalistics* and the Adams's colonial adaptation, *Bertillonage* and fingerprinting were used in efforts to maintain the view that nomadic peoples were criminals.

The problem for policing in the 1920s and 1930s was how to incorporate these and the burgeoning interest in 'scientific aids' into efficient detective and policing mechanisms. In particular, how would the development and professionalization of the role of detective be achieved to make scientific detection a realistic aim? It is to these questions that the next chapter turns.

Notes

1 J. E. Early, 'Technology, modernity, and "the little man": Crippen's capture by wireless', *Victorian Studies*, 1996, 39 (3): 309–337.
2 K. D. Watson, *Dr Crippen*, Kew, Richmond: The National Archives, 2007.
3 R. McWilliam, *The Tichborne Claimant. A Victorian Sensation*, London: Hambledon Continuum, 2007.
4 H.T.F. Rhodes, *Alphonse Bertillon: Father of Scientific Detection*, London: George G. Harrap & Co., 1956.
5 S.A. Cole, *Suspect Identities: A History of Fingerprinting and Criminal Identification*, Cambridge, MA: Harvard University Press, 2001.
6 D. Edgerton, 'The linear model did not exist', in K. Grandin, N. Worms, and S. Widmalm (eds), *The Science-Industry Nexus: History, Policy, Implications*, Sagamore Beach, MA: Science History Publications, 2004, 31–57; B. Godin, 'The linear model of innovation: The historical construction of an analytical framework', *Science Technology and Human Values*, 2006, 31 (6): 639–667.

7 B. Latour, *Science in Action*, Cambridge, MA: Harvard University Press, 1987, pp. 174–175; K. Asdal, B. Brenna, and I. B. Moser, (eds), *Technoscience: The Politics of Interventions*, Oslo: Unipub, Oslo Academic Press, 2007.
8 R. Bud, 'Making sense of modernity: The categories of pure and applied science in the public sphere of early twentieth-century Britain'. Paper presented at the British Society for the History of Science annual conference, St Andrews, UK, 2014.
9 R. Bud, 'Framed in the public sphere: Tools for the conceptual history of "applied science" – a review paper', *History of Science*, 2013, 51: 413–433.
10 Ibid.
11 Bud, 'Making sense'.
12 F. W. Taylor, *Scientific Management: Comprising Shop Management, the Principles of Scientific Management, Testimony Before the Special House Committee*, New York: Harper, 1911.
13 C.R.M. Cuthbert, *Science and the Detection of Crime*, London: Hutchinson, 1958; L. C. Nickolls, *The Scientific Investigation of Crime*, London: Butterworth, 1956; H. J. Walls, *Expert Witness: My Thirty Years in Forensic Science*, London: John Long, 1972.
14 D. Edgerton, *The Shock of the Old: Technology and Global History since 1900*, London: Profile, 2006.
15 E.g., see W. Roy, 'Traffic control in Edinburgh', *The Police Journal*, 1930, 3 (2): 262–276.
16 For police boxes see F. J. Crawley, 'Decentralization and the police box system', *The Police Journal*, 1928, 1 (1): 118–127.
17 J. Maxwell, 'The English police system: General developments and outstanding features', in L. Radzinowicz and J.W.C. Turner (eds), *Penal Reform in England*, London: Macmillan, 1946, 60–97, p. 87.
18 For moral panics, see C. Krinsky, *The Ashgate Research Companion to Moral Panics*, Farnham and Burlington, VT: Ashgate, 2013. For concerns about technologies being used for nefarious ends, see C. Marvin, *When Old Technologies Were New: Thinking About Electric Communication in the Late Nineteenth Century*, New York and Oxford: Oxford University Press, 1988.
19 J. Wajcman, *Feminism Confronts Technology*, Cambridge: Polity, 1991.
20 For dystopian Internet and computer use, see A. Adam, *Artificial Knowing: Gender and the Thinking Machine*, London and New York: Routledge, 1998; A. Adam, *Gender, Ethics and Information Technology*, Houndmills, Basingstoke: Palgrave Macmillan, 2004. For the introduction of the telegraph, see T. Standage, *The Victorian Internet: The Remarkable Story of the Telegraph and the Nineteenth Century's Online Pioneers*, London: Phoenix, 1999.
21 Marvin, *When Old Technologies Were New*, p. 92.
22 H. Machado and B. Prainsack, *Tracing Technologies: Prisoners' Views in the Era of CSI*, Fanham and Burlington, VT: Ashgate, 2012, p. 57.
23 E. Beauregard and M. Bouchard, 'Cleaning up your act: Forensic awareness as a detection avoidance strategy', *Journal of Criminal Justice*, 2010, 38 (6): 1160–1166.
24 Lucas, *Forensic Chemistry*, 1921, p. 15.
25 P. Knepper, *The Invention of International Crime: A Global Issue in the Making, 1881–1914*, Basingstoke and New York: Palgrave Macmillan, 2010, p. 3.
26 S. Karstedt, 'Strangers, mobilisation and the production of weak ties: Railway traffic and violence in nineteenth-century South-West Germany', in B. S. Godfrey, C. Emsley and G. Dunstall (eds), *Comparative Histories of Crime*, Devon: Willan, 2003, 89–109.

27 Knepper, *The Invention*, p. 3.
28 Ibid.
29 K. Laybourn, and D. Taylor, *Policing in England and Wales, 1918–39: The Fed, Flying Squads and Forensics*, Basingstoke and New York: Palgrave Macmillan, 2011, p. 186.
30 For banditry in general, see E. J. Hobsbawm, *Bandits*, Harmondsworth: Pelican 1972; G. Seal, 'The Robin Hood principle: Folklore, history and the social bandit', *Journal of Folklore Research*, 2009, 46 (1): 67–89. For banditry in colonial India, see K. A. Wagner, *Thuggee: Banditry and the British in Early Nineteenth-Century India*, Houndmills, Basingstoke: Palgrave Macmillan, 2007.
31 K. Laybourn, and D. Taylor, *Policing in England and Wales, 1918–39: The Fed, Flying Squads and Forensics*, Basingstoke and New York: Palgrave Macmillan, 2011, p. 186.
32 Ibid.
33 Ibid., pp. 189, 202–203.
34 T. Cullen, *Crippen: The Mild Murderer*, London: Bodley Head, 1977.
35 W.J. Dew, *I Caught Crippen: Memoirs of Ex-Chief Inspector Walter Dew, C.I.D. of Scotland Yard*, London and Glasgow: Blackie & Son Ltd, 1938.
36 For descriptions of modern DNA analysis of skin from a slide in the Crippen case, see D. R. Foran, B. E. Wills, B. M. Kiley, C. B. Jackson and J. H. Trestrail III, 'The conviction of Dr. Crippen: New forensic findings in a century-old murder', *Journal of Forensic Sciences*, 2011, 56 (1): 233–240.
37 For Crippen's description reported in a contemporary newspaper, see 'The mystery of Hilldrop-Crescent', *Daily Mail* Friday, 15 July, 1910, p. 5. For Bertillon and standardized description, see Cole, *Suspect Identities*, p. 45.
38 'The mystery of Hilldrop-Crescent', *Daily Mail* p. 5.
39 Early, 'Technology', p. 312.
40 Ibid. pp. 318–319.
41 Dew, *I Caught Crippen.*
42 Yellon Albion Magazine Oct. 1909, quoted in *Daily Mail* 16 July 1910. Also see Early, 'Technology', p. 318.
43 Early, 'Technology'.
44 C. Lombroso, *Criminal Man*. Translated by M. Gibson and N. H. Rafter. Durham, NC: Duke University Press, 2006. (original publication date 1876)
45 Quoted in Cullen, *Crippen* p. 15.
46 Early, 'Technology', p. 319.
47 Ibid.
48 *News of the World*, 7 August 1910 quoted in Early, 'Technology', p. 314.
49 Early, 'Technology'.
50 Dew, *I Caught Crippen*, p. 40.
51 Early, 'Technology', p. 309.
52 T. Campbell, *Wireless Writing in the Age of Marconi*, Minneapolis: University of Minnesota Press, 2006.
53 Early, 'Technology'.
54 'Wireless stations on the way to Quebec', reproduced in J. Goodman, *The Crippen File*, London and New York: Allison & Busby, 1985, p. 29. I have been unable to locate the original source of this diagram.
55 Dew, *I Caught Crippen.*
56 Early, '*Technology'*, p. 315.
57 *The New York Times*, 27 July 1910 quoted in Early, '*Technology*', p. 327.
58 Early, '*Technology*', p. 327.

59 'Race over ocean to catch Crippen', *The New York Times*, 24 July, 1910, p. 3.
60 'Crippen and girl on the Montrose', *The New York Times*, 29 July, 1910, p. 1.
61 J. Rowbotham, K. Stevenson and S. Pegg, *Crime News in Modern Britain: Press Reporting and Responsibility: 1820–2010*, Houndmills, Basingstoke: Palgrave Macmillan, 2013, p. 84.
62 J. Rowbotham and K. Stevenson (eds), *Criminal Conversations: Victorian Crimes, Social Panic and Moral Outrage*, Columbus: Ohio State University Press, 2005, Introduction p. xxvi.
63 Ibid., p. 106.
64 Ibid., p. 107. See Goodman, *The Crippen File* for contemporary press cuttings relating to the Crippen case.
65 E. Larson, *Thunderstruck*, New York: Broadway Books, 2007, p. 379.
66 H. Shpayer-Makov, *The Ascent of the Detective: Police Sleuths in Victorian and Edwardian England*, Oxford and New York: Oxford University Press, 2011, p. 218.
67 Early, 'Technology', p. 309.
68 'The long arm of "wireless" ', *Daily Mail*, 25 July 1910, p. 6.
69 Ibid., p. 7.
70 'Science and the fugitive', *Daily Mail*, 30 July 1910.
71 Ibid.
72 Ibid.
73 Ibid.
74 B. Newhall, History of photography: Photojournalism, *Encyclopedia Britannica*, 2014, available at http://www.britannica.com/EBchecked/topic/457919/history-of-photography/252857/Photojournalism, accessed 27 August 2014.
75 M. McFarland, *Chronology of Key Events in the History of the Daily Mail*, 2013, available at http://gdc.gale.com/assets/files/daily_mail/chronology_of_notable_events.pdf, accessed 27 August 2014.
76 F. A. MacKenzie, *The Mystery of the Daily Mail: 1896–1921*, London: Associated Newspapers, 1921, pp. 55–56.
77 'Science and the fugitive', *Daily Mail*, 30 July 1910.
78 T. Thorne Baker, *The Telegraphic Transmission of Photographs*, New York: Van Nostrand, 1910, p. v.
79 Ibid., p. 127–128.
80 Ibid., pp. 137–138.
81 Ibid., p. 141.
82 K. Watson, *Poisoned Lives: English Poisoners and Their Victims*, Hambledon and London: London and New York, 2004, pp. 52–53.
83 Ibid, p. 65.
84 P.H.A. Willcox, *The Detective-Physician: The Life and Work of Sir William Willcox*, London: Wm Heinemann, 1970, pp. 26–27.
85 Watson, *Poisoned Lives*, pp. 55–56.
86 A. Rose, *Lethal Witness: Sir Bernard Spilsbury Honorary Pathologist*, Chalfont, Stroud: Sutton, 2007, pp. 22–30.
87 I. Burney, and N. Pemberton, 'The rise and fall of celebrity pathology', *British Medical Journal*, 2010, 341: 1319–1321.
88 Watson, *Dr Crippen*, p. 44.
89 Lucas, *Forensic Chemistry*, p. 227.
90 Ibid., p. 228,
91 Ibid., p. 229.

92 Watson, *Dr Crippen*, p. 105; Goodman, *The Crippen File*, p. 89; Letter from H. G. Stevenson Coppin to Sir William Willcox, 5 January, 1925; Letter from Sir William Willcox to H. G. Stevenson Coppin, 9 January 1925.
93 J. Caplan, and J. Torpey, 'Introduction', in J. Caplan and J. Torpey (eds), *Documenting Individual Identity: The Development of State Practices in the Modern World*, Princeton and Oxford: Princeton University Press, 2001, 1–13, p. 1.
94 McWilliam, *The Tichborne Claimant*.
95 E. Higgs, *Identifying the English: A History of Personal Identification 1500 to the Present*, London and New York: Continuum, 2011, p. 123.
96 Ibid., p. 125.
97 M. Kaluszynski, 'Republican identity: Bertillonage as government technique', in J. Caplan and J. Torpey (eds), *Documenting Individual Identity: The Development of State Practices in the Modern World*, Princeton and Oxford, Princeton University Press, 2001, 123–138, p. 123.
98 Ibid., pp. 123–124.
99 Ibid., p. 124.
100 Ibid., p. 128.
101 Cole, *Suspect Identities*; Kaluszynski, 'Republican identity'; Rhodes, *Alphonse Bertillon*.
102 Rhodes, *Alphonse Bertillon*, p. 199.
103 Ibid. The Dreyfus affair was a political scandal in France running from 1894 to 1906 centring on the question of whether Jewish army captain Alfred Dreyfus sold military secrets to the Germans. Questions of anti-Semitic sentiments were raised. Bertillon was a prosecution witness who testified that the handwriting in question was Dreyfus's. He was testifying well beyond his sphere of expertise. The questioned document was eventually found to be a forgery. See 'Dreyfus affair', *Encyclopaedia Britannica*, available at http://www.britannica.com/EBchecked/topic/171538/Dreyfus-affair, accessed 5 May 2015.
104 Cole, *Suspect Identities*, p. 34.
105 Rhodes, *Alphonse Bertillon*, p. 75.
106 D. Bloor, *Knowledge and Social Imagery*, London: Routledge & Kegan Paul, 1976; W. E. Bijker, T. P. Hughes, and T. J. Pinch (eds), *The Social Construction of Technological Systems: New Directions in the Sociology and History of Technology*, Cambridge, MA: MIT Press, 1987.
107 Cole, *Suspect Identities*.
108 Ibid., p. 36.
109 Ibid., p. 43.
110 H. M. Collins, *Tacit and Explicit Knowledge*, Chicago and London: University of Chicago Press, 2010.
111 H. M. Collins, 'The TEA set: Tacit knowledge and scientific networks', *Science Studies*, 1974, 4: 165–186.
112 Research on the design of technological systems within science and technology studies has described the way in which certain types of users or certain beliefs can be 'inscribed' in the design of technological systems. See M. Akrich, 'The De-Scription of Technical Objects', in W. E. Bijker and J. Law (eds), *Shaping Technology/Building Society: Studies in Sociotechnical Change*, Cambridge, MA: MIT Press, 1992, 205–224 for a description of inscription.
113 M. M. Astrahan, and D. D. Chamberlin, 'Implementation of a structured English query language', *Communications of the ACM*, 1975, 18 (10): 580–588.

114 R. B. Du Boff, 'The rise of communications regulation: The telegraph industry, 1844–1880', *Journal of Communication*, 1984, 34 (3): 52–66; Standage, *The Victorian Internet*.
115 Cole, *Suspect Identities*, p. 45.
116 Ibid.
117 Ibid., p. 51.
118 Akrich, 'The de-scription'.
119 Kaluszynski, 'Republican identity', p. 127.
120 Rhodes, *Alphonse Bertillon*, p. 107.
121 H. Söderman, 'Science and criminal investigation', *Annals of the American Academy of Political and Social Science*, 1929, 146 (1): 237–248, pp. 239–240.
122 Ibid., p. 240.
123 Kaluszynski, 'Republican identity', p. 129.
124 Ibid., p. 131.
125 Ibid., p. 132.
126 Ibid., pp. 136–137.
127 Galton used the term 'little worlds' for fingerprints in: F. Galton, *Finger Prints*, London and New York: Macmillan, 1892, p. 2.
128 Cole, *Suspect Identities*.
129 Ibid:, C. Sengoopta, *Imprint of the Raj: How Fingerprinting Was Born in Colonial India*, Basingstoke and Oxford: Pan, 2003.
130 Cole, *Suspect Identities*, p. 60.
131 Sengoopta, *Imprint of the Raj*, pp. 41–43.
132 Ibid., p. 42.
133 Ibid., p. 48.
134 Ibid., p. 51.
135 Sir William Herschel from 1871.
136 Sengoopta, *Imprint of the Raj*, p. 57.
137 Ibid., p. 72.
138 Ibid., pp. 73–74.
139 Ibid., p. 78.
140 From W. J. Herschel, *The Origin of Finger-Printing*, London: Humphrey Milford/Oxford University Press, 1916, p. 30. Herschel recorded his prints at age 26, 44 and 83.
141 Sengoopta, *Imprint of the Raj*, p. 79.
142 Ibid., p. 80.
143 H. Faulds, 'On the skin-furrows of the hand', *Nature*, 1880, 22 (October 28): 605.
144 Sengoopta, *Imprint of the Raj*, p. 87.
145 Ibid., p. 88.
146 F. Galton, *Hereditary Genius: An Inquiry into Its Laws and Consequences*, London: Macmillan, 1869.
147 Galton, *Finger Prints*.
148 J. Wayman, A. Jain, D. Maltoni, and D. Maio, D. (eds), *Biometric Systems: Technology, Design and Performance Evaluation*, London: Springer-Verlag, 2005, p. 24.
149 Ibid.
150 Sengoopta, *Imprint of the Raj*, p. 121.
151 Ibid., p. 126.
152 Galton, *Finger Prints*.
153 See F. R. Cherill, *The Fingerprint System of Scotland Yard: A Practical Treatise on Finger Print Identification for the Use of Students and Experts and A Guide*

for Investigators when Dealing with Imprints Left at the Scenes of Crime, London: HMSO: 1954, especially Introduction pp. 1–12.

154 Sengoopta, *Imprint of the Raj*, p. 141.

155 Ibid., pp. 143–144.

156 Wayman et al., *Biometric Systems*, p. 24.

157 For a description of the Henry system, see Ibid., p. 25, and Sengoopta, *Imprint of the Raj*, p. 213.

158 Wayman et al., *Biometric Systems*, p. 25.

159 Ibid., p. 27.

160 Ibid., pp. 26–28.

4 Scientific detection, scientific aids and forensic science laboratories

From the 1880s, European and American police forces began to look to the Bertillon's anthropometric system as a suitable model for scientific policing[1]; it had the reputation for being 'ploddingly bureaucratic yet devastatingly effective'.[2] However, at least in the UK, the spur towards the development of scientific policing came a little later, in the early years of the twentieth century and encompassed a range of potential models. Although the efficient standard of organization offered by the Bertillon system clearly appealed, certainly in its promise of a viable record keeping system, less so in its intricate measuring requirements, it was but one building block in the edifice of scientific policing. In the UK, effectiveness in the application of technoscience to policing and detection rested on a mix of 'plodding bureaucracy' and 'the scientific use of the imagination', a phrase made popular by the Victorian physicist and popular lecturer, John Tyndall, and quoted by Sherlock Holmes in *The Hound of the Baskervilles* to describe his art of scientific detection.[3] Indeed, the stage was set for both bureaucracy *and* imaginative use of technoscience as official committees debated the best approaches to criminal identification, detection and the introduction of scientific techniques, while individual police officers and scientists suggested novel approaches towards the division of labour between police officer and scientist in the development of scientific aids to detection.

In 1893, the UK government's Home Office appointed a committee to compare methods of identifying recidivists and to recommend the best approach towards identification. Mindful that whatever the chosen method, or methods, it had to be relatively easy for measurements to be taken by police and prison officers, the Troup Committee (1894) reported the following year, recommending a combination of anthropometric measurements and fingerprinting.[4] Only a few years later the Belper Committee (1901) recommended that the criminal identification system should be based on fingerprinting alone.[5] During its deliberations, members of the Troup Committee travelled to Paris to see Bertillon's system and were impressed by its record system. '[A]nthropometry's strongest selling point lay not in the proof of identity, but rather in the simple and effective system of classification, which Galton's system of fingerprinting lacked in 1893.'[6] Fingerprinting did not yet encompass a classification system which was at the same stage of maturity and effectiveness, although this was just over the horizon; hence it was not at that point recommended as a single method of criminal identification.[7] It did not yet have the technoscientific wraparound in terms of a database and search technology to make it an immediately viable forensic technique. In any case, the question of matching a latent print found at a crime scene to an individual suspect remained an issue.

Over the next few years the anthropometric system came to be seen as increasingly problematic, partly because different operators often obtained different measurements from the same person, but also because the instrumentation and the metric system embodied in the instruments were quite literally foreign to British operators. In Britain, the *Bertillonage* star ascended at the end of the nineteenth century only to set in the early twentieth century. The Belper Committee recommended the low cost and simplicity of fingerprinting technology, emphasizing the persistence of fingerprints throughout life which made it a system usable on juveniles as well as adults.[8] Although the committee was not able to see the Henry system of classification in operation in Bengal, nevertheless the imminent availability of a suitable classificatory system must have been a positive force in convincing the British police authorities to adopt fingerprinting as their main means of criminal identification.[9]

Division of labour is a central theme for the introduction of scientific detection, scientific aids and forensic science. It was not only a question of what to do; it was a question of who would do what. Just as with other forms of technoscience employed in relation to crime detection, the protocols needed to be worked out and agreed to, and several suggestions and models of working emerged, particularly in the 1930s when science in British policing began to come of age. Such protocols not only defined professional roles, they were also significant in epistemological terms; they made disciplines and knowledge.

Fingerprinting was part of what the later Detective Committee (HMSO 1938) called 'police technique' or police science, to be undertaken in house by local police forces. It was not a technique that was regarded as requiring the services of specialist laboratories.[10] However, records had to be held centrally; a central, searchable data store was vital for the system to be effective in the identification of repeat offenders. A fingerprint department was established in 1901 in New Scotland Yard as part of the Metropolitan Police's CID and Henry became the new Assistant Commissioner in charge of the CID and fingerprint department.[11] Detectives examined and interpreted the fingerprints collected by police officers in local forces. So fingerprinting as a police technique, with a clear division of labour and an appropriate central database, was working effectively in England and Wales by the 1930s when other scientific aids were gradually coming into the spotlight.

The Detective Committee's definition of police technique included photography, processes for developing latent fingerprints, taking casts of footprints or coarse tool marks and appropriate methods for searching and for handling and packing trace evidence; the committee recommended that all these should be in the detective training syllabus.[12] Of course, the division between police technique and forensic science, which was the delineation between what the police did and the province of the laboratory scientist, was yet to achieve a clear boundary.

Decisions had to be made about what fell to which official or scientist. Generally, administrative closure and the perceived simplicity and frequency of use of a technique marked it as falling under the heading of police technique. However, there was sometimes no *ab initio* reason why a procedure should be categorized thus, and expert scientists of a forensic bent often took a great deal of interest in what seemed like the prosaic aspects of police science, a prime example being the packing of crime scene materials.[13]

Although the Troup and Belper committees were commissioned to address the perennially troubling matter of accurately identifying criminals, they were also a central part of the story of efficient use of effort and establishment of authority in policing and detection and the adoption of technoscientific approaches to underpin that authority.[14] The impetus towards better, centrally organized police forces and a growing realization that science could be directly employed in the detection of crime were major driving forces towards an approach to policing which was regarded as appropriately modern and scientific. Both efforts were championed by senior figures in the police and the Home Office, but there was also considerable interest in scientific methods, if not quite at a grassroots level, then at least amongst the working ranks of police and detectives, as well as amongst some academics and independent scientific expert witnesses. Cyril Cuthbert was the archetype of the policeman-scientist, one of the few who was able to develop a professional identity based on the intersection of policing and science.[15]

'Science' was interpreted widely to encompass the technoscientific developments alluded to in the previous chapter, such as the telegraph, radio and automobiles, and a more generalized 'scientific approach' which included appropriate records management and extensively detailed discussions of identifying, handling and packing trace materials obtained from crime scenes, the sorts of things which would be badged under the heading of police technique if they were not already, alongside what we would recognize as specific scientific analyses and techniques employed in the fight against crime.[16] The scientific approach was encapsulated in the 'scientific aids' movement which was the British response to criminalistics. 'Scientific aids' offered a kind of crime scene management front end for police officers and detectives to the specific forensic scientific analyses which were developing and being used in laboratories. The champions of the 'scientific aids' movement contributed to attempts to reform policing, helping to establish the detective role on a firm footing with appropriate scientific support. This group included scientists such as F. G. Tryhorn (1893–1972) and senior civil servants, in particular, Arthur Dixon (1881–1969), whose role is described later. Senior police officers serving on the various committees, who wrote about the potential of science in the fight against crime in the pages of the *Police Journal* and in other periodicals or books, were also part of the impetus to put detection on a scientific basis.[17]

This chapter considers how scientific policing and detection developed in the UK in the first part of the twentieth century, bearing in mind the tensions between 'scientific' as a metaphor for a modern, progressive organization, at least in the eyes of those in administratively powerful positions, and the concerns of grassroots police officers as to what would be expected of them in relation to the demands of the new scientific approach. Policing and police reforms in the wake of the First World War relied to a considerable extent on the rhetoric of science and technology as modernizing forces. However, there was hardly unalloyed enthusiasm for the introduction of new technologies and new scientific techniques. In fact, there was considerable ambivalence towards the introduction of science into policing and detection, not least amongst rank-and-file detectives and police officers, and this ambivalence was played out in the initial reception of the new Metropolitan Police Laboratory in the second half of the 1930s, and to a lesser extent, towards the opening of the North Western Laboratory in 1938.[18] 'Scientific' could, of course, stand for desirable, progressive, modern policing, but it could also stand for untried and possibly untrustworthy technologies, which were not always readily assimilated as reliable substitutes for traditional policing knowledge and methods.

For instance, lest we imagine that the public and the police demonstrated unbridled enthusiasm for new sciences and technologies, we should note the question of whether a technological device or a scientific analysis was to be trusted was often raised. Despite the drama of the Crippen arrest and the much-vaunted role of the wireless telegraph in its successful execution, the police rarely cabled ahead to apprehend a suspect escaping in a steamship. Instead, they usually acquired a warrant and sent a detective in pursuit which would, at least, go some way towards explaining why Dew set off after Crippen and Le Neve on a faster ship rather than cabling the Canadian authorities.[19]

Scientific policing – an age of anxiety

The question of trusting technology was also reflected in anxieties commonly expressed about the dehumanizing effects of science and the difficulties of understanding technological developments. As the final chapter describes, there was considerable enthusiasm for science in crime detection in fiction, but it was safe in fictional form when one did not have to worry about the consequences. Police officials were no less prey than the public to worries over the potential of new scientific and technological developments and the effects these might have on their working lives. People marvelled at new technological wonders but, at the same time, worried about their negative potential.[20]

Such fears were one matter, but outright hostility was another. Al Dunlap, editor of a popular US police magazine *The Detective*, captured the mood in eloquent style in an article entitled 'Science versus practical common sense'.[21] It was very clear that he tacked his colours to the mast of common sense. Referring to the controversy raging in criminal justice quarters between supporters of science and old-style detectives, Dunlap quoted the clash between Captain Leonard of the Michigan State Police, a supporter of science in policing, and Captain Mathewson, Chief of Detectives at the San Francisco Police Department, who clearly was not, at a Police Chiefs' Convention in 1931.[22] Mathewson declared that what had been said about college-educated police and detectives was 'all bunk'. Given an experienced practical detective with ten men, he could do much more than a supposed expert with a hundred experts in tow. Mathewson did not pull his punches, declaring that so-called experts '[c]ouldn't put a harness on a mule, let alone catch a crook'.[23] There were too many people who regarded themselves as experts, and they had a credulous public believing 'that they are so scientific that the crooks would respond to engraved invitations to visit police headquarters and surrender'.[24] It was not just a question of police officers being asked to develop new skills; it was also that such skills, particularly when in the hands of those who were supposedly better educated, threatened to supplant traditional policing knowledge. Dunlap may have been expressing his pain at the rather weighty chip he was carrying on his shoulder; although senior British police officers might not have expressed themselves quite so picturesquely, they were hardly immune to similar sentiments.

Concern over new scientific techniques centred on the question of police training. Until police officers and detectives knew what could be achieved by scientific methods, they would be less likely to pass material to the new laboratories for analysis. Hamish Walls, the Metropolitan Laboratory's first physicist, bemoaned the lack of material being sent to the laboratory from the Metropolitan Police in the laboratory's early years.[25] The comprehensive training, including training on scientific aids, detailed in the Report of the Detective Committee (HMSO 1938) offered a potential means to mitigate ignorance of scientific methods, but this did not mean that change would quickly be forthcoming.

Apprehension must be set against the almost-unqualified enthusiasm for scientific and technological advances, especially when used against crime, displayed in the pages of forward-looking periodicals such as the *Police Journal*, a police service journal which began publication in 1928 and in whose pages many of the developments of the 'scientific aids' movement in policing would be described. The introduction of science into crime detection is inextricably linked to organizational matters, and anxiety about organizational changes brought about by new technologies was firmly connected to anxiety about scientific and technological innovations. We did not invent the problem of the impact of new technology on jobs at the end of the twentieth century; it is a perennial concern.[26]

The systematic use of scientific techniques was barely feasible without a system of criminal record keeping and sharing and without suitable communications. Police forces internationally had little capacity to share information, given that their record keeping was inadequate. As Chapter 3 discussed, criminals could be as good as, if not better than, the police at availing themselves of new technologies, including transport and communications technologies.[27] It was relatively easy for criminals to disguise themselves and melt into the crowd at a time when records were neither centralized nor was information widely shared.

The Desborough Committee – the beginnings of police reorganization

At the beginning of the First World War, the police force was viewed as 'still a collection of Victorian bric-à-brac'.[28] At the end of the war the situation was no better. Police forces were under strength, there was concern about the physical fitness of recruits, small police forces proliferated and police practice was locally driven and was far from standard across England and Wales.[29] Hence police organization in England and Wales was poorly equipped to take advantage of the promise of new scientific and technological advances in solving crime, particularly those advances which involved specialist skill, equipment, sharing of information and the ability to operate large centralized databases of criminal information as demanded by systems such as fingerprinting and *Bertillonage*.

A crisis provided the spur for considering ways to improve police organization – the two police strikes of 1918 and 1919.[30] These were regarded with considerable alarm, and it must have seemed to some to be the beginning of the socialist revolution in the UK. The inter-war years were turbulent times in Britain, and there was much concern over the rise of political radicalism and anything that seemed to herald extremism. Against this backcloth, the British government's attempts to centralize and organize policing were not just efforts to modernize and to make policing more efficient; they were also a means to ensure that the police would have appropriate pay, conditions and training so that they would resist the influence of extremist political groups.[31] The adoption of new science and technology in police work gradually came to be regarded as part of this process; although this was somewhat under the surface, scientific aids in crime detection as tools of the establishment in creating and maintaining order therefore embodied a political dimension. Science and technology were not just being mobilized to control criminals, but were also being organized to discipline those who fought crime as part of a Foucauldian disciplinary regime.[32] Scientific approaches as agents of superior organization and crime detection were part of the machinery of the state in combatting political extremism amongst its own law enforcement officers.

The deliberations of the committee presided over by Lord Desborough formed the basis of the Police Act of 1919 which set central pay and conditions and set up the Police Federation to replace the Police Union which was banned because of its role in organizing the police strikes.[33] The Desborough Committee, although not regarded as particularly successful, is generally viewed as heralding the beginning of the modernization of the UK police, not just in terms of pay and conditions, but also in terms of centralization of organization and resources which would pave the way in due course for the introduction of scientific aids into policing. One role of the committee was to consolidate the involvement of the Home Office in policing to ensure uniformity and cooperation over the forces and to enforce proper police administration.[34] Arthur Dixon, secretary to the Desborough Committee, was an enthusiastic centralizer of the police service and would become the prime driver in the 'scientific aids' movement.[35] He was to be a central figure on the Detective Committee in the 1930s and held a pivotal role in the creation of the new forensic science laboratory network.

Professionalizing the detective role – Home Office committees

The importance of professionalization of detection should not be underestimated in the development of the new forensic sciences. Arguably, the development of the detective role was key to linking laboratory forensic science definitively to the repertoire of modern policing and detection, alongside fingerprinting and other police techniques and developments in communications and transport technology. But histories of policing and detection often only touch on forensic science lightly; it is usually only included as a small part of the story.[36] Similarly, histories of forensic medicine (given that there have been so few academic histories of forensic science) do not generally put the role of the police, and organizational matters, centre stage.

Detective activity in policing has a long history, but the role of the detective as a professional position separate from the ordinary police officer began to develop towards the end of the nineteenth century in the UK.[37] In the Victorian and Edwardian eras detectives were recruited internally from the ranks of uniformed policemen, where they were expected to serve their time before being selected for the role of detective. Whereas policing strategy usually emphasized physical strength and a suitably commanding presence for the position of rank-and-file constable, by contrast, intelligence and judgment were regarded as the ideal attributes for the detective.[38] Literacy skills and education were seen as increasingly important, however, as detectives were recruited from uniformed officers and, as many of these had originally worked in manual jobs, they were not usually drawn from well-educated echelons of society.[39] Lack of scientific education would certainly have contributed to anxieties about the demands that the new scientific aids would make of the police officer and detective. Hence, there was a tension between the high expectation of the skills of those selected for the detective role and the talents of the forces from which such detectives would inevitably be selected.

The possibility that science would triumph over crime still appeared a distant dream, and police forces were slow in becoming organized and professionalized. In England and Wales in the period after the Great War, there were no forensic laboratories on the scale of Continental forensic laboratories such as Locard's world-renowned Lyon laboratory.[40] Continental forensic laboratories were often held up as exemplars of well-oiled crime fighting machines, in contrast to the sorry state of affairs in England. Until the early 1930s, forensic activity in English police forces largely depended on the activity of a local force in whatever laboratory provisions it might make, coupled with the strength of detective activity as a balance to traditional preventive policing. Detective and laboratory facilities were very variable. For instance, in the 1930s, the Bradford force boasted a detective force, links with the local college to provide infra-red and ultra-violet analyses of evidence and pioneered colour photography in police work, whereas, by contrast, in 1940 the new Chief Constable of the Oxfordshire police apparently arrived at his new post to find no specialist sections and no CID.[41]

The Desborough Committee did not achieve the desired increases in efficiency. A decade later, Dixon reported that there were 58 county forces and 121 borough forces in England, including the Metropolitan Police and a separate City of London Police. Historically, the latter had been fairly small. Of these forces, nearly 100 had fewer than 100 officers, 'a source of weakness in the system'.[42] Where small, local forces proliferated, the number of trained detectives remained low. Of the 35,000 to 36,000 strong county and borough forces in the same period, there were 1780 detectives, implying fewer than 10 detectives per force. The percentage of detectives amounted to less than 5 per cent of the total police strength of England and Wales.[43]

So by the beginning of the 1930s policing had still not been sufficiently centralized and streamlined, crimes against property continued to be of major public concern and CID strength was seriously lacking.[44] Against the backcloth of rising crime, *ad hoc* or absent laboratory arrangements and a fragmented police system, where detective forces were still under strength, the work of two Home Office committees convened in the 1930s is highly relevant. The first was the Advisory Committee on the Scientific Investigation of Crime (HMSO 1936), or what was termed the 'Scientific Advisory Committee'. The second was the Departmental Committee on Detective Work for England and Wales (HMSO 1938), or, as it was called, the 'Detective Committee'.[45] On the face of it, one might have expected the former to be of more significance to the introduction of forensic science to policing but this was not the case; the Report of the Scientific Advisory Committee did not do what it said on the tin. One of its remits was to oversee the development of the new Metropolitan Police Laboratory which opened its doors in 1935. The report noted that good progress had been made on this front (a claim which would look increasingly doubtful as the decade progressed) and recommended an increase in staff on the assumption that demands on the laboratory were bound to grow as police became more knowledgeable about the application of science to crime investigation. This was the only recommendation of the report which was accepted. The remainder of the report was given over to arguments for the inauguration of a national medico-legal institute.[46] The Scientific Advisory Committee was dominated by senior medical men and chaired by Lord Trenchard, Commissioner for the Metropolitan Police, who neither gave the committee a strong steer nor understood that there was a difference between forensic medicine and forensic science.[47] The proposed central medico-legal institute (which was never built) would have been expensive and would not have addressed Home Office concerns about crimes against property, as its focus would have been murders and sudden deaths.[48] However useful it would have been to build it, had the money been forthcoming, it was not an institution which would have driven forward the setup of a network of regional forensic science laboratories. The Scientific Advisory Committee, although recognizing that the new Metropolitan Laboratory had to develop close links with police institutions, had little to say about how this was to be achieved.[49] Its concerns centred firmly on arguments for building the proposed medico-legal institute.

The report of the Scientific Advisory Committee argued that, although there was immense value in developing a system of laboratories to bring science to bear on crime: 'This development will, however, only solve part of a larger problem and to deal comprehensively with the subject it is necessary to go beyond the restricted field of purely police requirements'.[50] Other countries had medico-legal institutes, and there were, of course, the influential chairs of forensic medicine at the universities of Edinburgh and Glasgow but there was nothing comparable in

London. The Metropolitan Laboratory could not be expected to handle the anticipated range of cases of industrial disease, post-mortem work and the academic development of forensic medicine. It is interesting that crime statistics, overwhelmingly pointing to the preponderance of crimes against property, had so little bearing on the committee's deliberations. The rest of the report focused on arguments for the development of such an institute. The role of science in detection of crime and how this could develop in relation to police institutions was not referred to subsequently. However, the report did signal that there was a dearth of specialist forensic organizational capacity all round, even though its focus was on the medical side. As we shall see in the next chapter, on occasion, forensic scientists expressed their frustrations with organizational arrangements relating to the medical aspects of forensic work. In any case, arguments for a medico-legal institute continued to rumble around for years in meetings of the Medico-Legal Society and in the pages of its journal, the *Medico-Legal and Criminological Review*.[51]

The Detective Committee

The Departmental Committee on Detective Work for England and Wales, a Home Office committee that operated between 1933 and 1938 and which submitted its report in 1938, the so-called 'Detective Committee', consisted mainly of senior police officers. In addition to three Home Office members, there was a clutch of senior police officers, including more than twenty chief constables.[52] In this respect the composition of the committee was significantly different from the membership of the Scientific Advisory Committee. The Detective Committee was appointed 'to inquire and report upon the organization and procedure of the police forces of England and Wales for the purpose of the detection of crime'.[53] Sub-Committee D (Scientific Aids) was chaired by R. M. Howe, Chief Constable of the Metropolitan Police. After promotion to Assistant Commissioner of the CID at Scotland Yard, Howe would go on to edit the fourth edition of *Criminal Investigation*, the English-language adaptation of Gross's *Handbuch* which indicated 'getting back to proper investigation' after the Second World War.[54]

The Detective Committee was set up to report on the detection of crime in England and Wales, specifically to enquire upon organization. The committee's work was divided into four related sections: selection and training of detectives, criminal records, communications and scientific aids in the detection of crime. This list is significant, as it links scientific analysis to the training of detectives and to other aspects of technology and records management in scientific detection, the whole spectrum of technoscience and information technologies. Such links were, of course, absent from the remit of the Scientific Advisory Committee.

The chief architect of the push for scientific aids through the 1920s and 1930s in the UK was Arthur Dixon, who chaired the committee.[55] As secretary to the earlier Desborough Committee, Dixon had a long history in campaigning for reforms to police administration; he was now Assistant Under-Secretary of State responsible for the Home Office Police Department. A career civil servant chiefly remembered for his work in introducing organizational efficiencies into the police and fire services in the inter-war years, he was also a keen amateur scientist. He wrote a popular book on atomic energy after he retired, reflecting his wider interests in the power of science, including the power of science in policing.[56] His combination of skills in high-level policing reforms and imaginative championing of science was to prove a winning combination in the introduction of scientific aids to detection.

The involvement of the Home Office in centralizing control over police organization and training and its role in professionalizing and growing the role of the detective, as well as its part in setting up the new laboratories, was a significant contribution to the creation of forensic science. It is not too far off the mark to claim that the Home Office had invented forensic science. Indeed, they may even have invented the term 'forensic science'.[57] Through the work of the Detective Committee in general and the ensuing inauguration of a forensic science laboratory network, the Home Office ensured the widespread use of the term by the end of the 1930s. This was coupled with the active promotion of 'scientific aids' through the work of the Detective Committee and through the publications of senior members of the 'scientific aids' movement, in particular the writings of Professor Tryhorn of University College, Hull, who was the senior scientist on the Detective Committee.[58]

Laybourn and Taylor argue that primary and secondary sources have tended to overlook the quantity of forensic work that was being undertaken in the 1930s in England and Wales.[59] However, even if this is the picture painted by standard criminological histories, some readily available primary and secondary sources confirm the scale of effort that went into setting up forensic science laboratories in the 1930s in tandem with forensic science analyses, which *were* being introduced in policing from the 1930s alongside the work of independent scientific expert witnesses and the continuing work of the public analysts in food and substance adulteration.[60]

It is important to consider the Detective Committee's report in some detail, as, not only did it reveal the current state of scientific detection in the UK in relation to record keeping, communications technology and scientific aids, it also looked to the future by supplying recommendations for the development of these areas and ways of achieving the required developments. Indeed, it was the manifesto of the 'scientific aids' movement. Such views were, of course, those of very senior members of English and Welsh police forces and senior civil servants. Nevertheless, training of police officers and detectives was addressed and the proposed content of training courses was spelled out at length. In the Report of the Detective Committee we find a clear link made between the role of the detective and the use of science, that is, forensic science, in the detection of crime, and the practical means of forging that crucial connection is reflected in the design of detective training courses. It did not just advocate attention to scientific aids – it specified that forensic science should be part of scientific detection.

Ostensibly only the last volume (Volume 5) of the Detective Committee's report was concerned with scientific aids, as such; nevertheless lengthy sections on crime records and communications (Volumes 2 and 3, respectively) reinforced the idea that the introduction of technoscience into crime detection involved record keeping and communications technology just as much as science. Volumes 1 and 2 reported on the police system of England and Wales and the selection and training of detectives, including a detailed description of the training that detectives should have in scientific aids and understanding scientific techniques. Volumes 2, 3 and 4 were not on sale to the public.[61]

Ambage suggests that the Detective Committee was successful (in contrast to the Scientific Advisory Committee and the earlier Desborough Committee) because it recommended things that were already happening.[62] In addition to this point, the fact that it spelled out requirements in considerable detail, effectively drawing up a shopping list, was helpful in breaking down its requirements into manageable, cost-effective chunks. Everything did not have to happen at once; indeed, the outbreak of war soon put paid to that idea. The committee adopted a different, and potentially more effective, strategy than pressing for building a large-scale and inevitably costly institution, such as the proposed medico-legal institution that was advocated by the Scientific Advisory Committee. Hence, the Detective Committee distributed its eggs of scientific detection across a number of baskets rather than putting them all into one big expensive basket. The report played an important role in restating and reinforcing troubling aspects of police organization and crime detection. It did no harm to remind the Home Office of these once again, and they were presented as a necessary backdrop to the reforms in the detective role the committee recommended. All the while the report made a strong link between modern detection and scientific aids. Chapter 1 of the report described the police system and the organization of Criminal Investigation (CID) or Detective Departments. The continuing large number of separate police forces was noted, as was the resultant effect on the ability of CID detectives to build up specialized skills, as some small forces had little opportunity to develop specialisms even though smaller forces could borrow detectives from the Metropolitan Police for serious crimes.[63] The central records and fingerprint systems provided by the Metropolitan Police were seen to be operating effectively. Despite these advances, the committee concluded that much still had to be achieved, including improved organization training, record keeping, communications and, of course, 'the application of Forensic Science to detective work in many classes of crimes'.[64]

The report reiterated the, by then, well understood observation that, in the 1930s, most crimes were against property rather than persons. Crimes of violence, including sexual offences, made up 2.5 per cent of indictable crime, violent crimes against property (burglary, shop and house breaking) amounted to 19.5 per cent, while non-violent theft was 75 per cent and currency and forgery crimes 1 per cent.[65] Crimes of burglary, house and shop breaking had tripled since the beginning of the Great War. Local versus national issues in crime detection were important. Greater movement of criminals reinforced the desirability of forces cooperating with one another in record keeping and communication. For the Detective Committee, the obvious advantages of centralization were to be achieved by maintaining adequate detective forces and equipment and centralizing record keeping, in the same way as the Fingerprint Registry and Criminal Record Office were maintained at New Scotland Yard.[66]

Volume 2 of the report dealt with the selection and training of detectives. An essential part of the training syllabus for new detective recruits involved instruction on what to do at the scene of a crime, including examining the scene, searching for 'clues'; handling and labelling items found at the crime scene and dealing with tracks and marks, photography and fingerprints.[67] The instructional course emphasized the practical, approaching the subject from the police officer's point of view and using exhibits from actual crimes as illustrations. A visit to a forensic science laboratory was encouraged during the course of instruction. The section on 'scientific aids' was to include 'explaining in general terms the various types of expert who may be of assistance in different types of cases'.[68] All members of the class (both those on the new recruits course and those on the longer detective training course) were to carry with them the 'Instructional Pamphlet on Scientific Aids' and were expected to refer to it in relation to the lectures within its scope.[69] This small, modest pamphlet was a key text in the scientific aids movement, as described later in this chapter.

A further course of instruction for detective training covered these topics in considerably more detail in an eight-week course, and it is notable, although not perhaps surprising, just how many forensic science topics were covered.[70] The preliminary lecture on scientific aids involved a fuller description of laboratory work, accuracy of scientific methods and the differences between scientific and other types of evidence. The evidential value of dust and dirt were explained with chemical methods of analysis and a description of a spectrograph. Guns and ballistics, fraud, forgery and handwriting were included, as was fingerprinting at length. The students were also expected to examine a prepared crime scene for scientific evidence and collect, note and pack material for sending to the expert. The Detective Committee Report is striking for its considerable detail and for its definition of the detective role, which it effectively defined through its comprehensive detective course. We can read the features of the detective training course as a job description for the role of the modern detective, and it is significant that it contains so much by way of reference to 'scientific aids' and scientific crime scene management, dealing with trace evidence, knowledge of scientific techniques, the techniques of fingerprinting and recording, as well as all the communications technology which the detective would require to know and use. The Detective Committee report clearly heralded the arrival of the scientific detective.

Sub-committee D – application of science to the investigation of crime

Volume 5 of the report, the responsibility of sub-committee D, indicated that a number of initiatives were in progress to improve the lack of scientific content in detective training courses.[71] The committee heard evidence from the forensic pathologist Spilsbury and Home Office Analyst Roche Lynch. Although both were medically trained (Roche Lynch had qualifications in medicine and chemistry), they were highly experienced in forensic work and both undertook chemical analyses. The Detective Committee reviewed work in the UK and overseas. Some forces in England and Wales had developed laboratories or took advantage of nearby laboratories and training. By the time the Detective Committee concluded its report, the Metropolitan Laboratory had been set up alongside the Metropolitan Police College at Hendon, and regional laboratories at Nottingham, Cardiff and Birmingham were already in operation. The instructional pamphlet on scientific aids had been issued to all forces by the Home Office and a series of Forensic Science circulars had also been initiated.[72] Apart from instruction at Hendon and Wakefield there had been other lectures in district detective conferences and other opportunities.

Although estimates of the volume of work were difficult to make, the committee stated its belief that around 4 per cent of indictable crimes would require scientific analysis of evidence, amounting to more than 10,000 crimes per year. There was clearly much scope for further development. The committee recognized that there were three grades of laboratory work: routine, specialist and research. To balance proposals for a medico-legal institute, the Detective Committee proposed its own institute to support the whole service in the form of a Forensic Science Centre which it was hoped would house a research laboratory, library, museum of specimens and the necessary administrative support for coordinating the work of the network of forensic science laboratories and disseminating information and liaising with universities. The proposed centre would also support the work of Home Office Forensic Science Adviser, which was a broad role designed to advise on the scientific work of the network of new laboratories. This was regarded as a more pressing requirement than a medico-legal institute for police purposes.[73] Inevitably, the central research and coordination facility did not materialize.

Perhaps mindful of the kinds of criticism made in the USA of 'college trained' detectives, the Detective Committee was anxious that the scientific work did not become too academic and far removed from activities required at the crime scene the way it sometimes did in other countries.[74] Knowing what to do about discovering and preserving materials from the crime scene was vital given that the person who arrived at the scene of the crime first was likely to be the uniformed constable. The report argued that both the police officer and the detective were required to know what could be achieved by the various laboratory techniques, and to that end they had to understand what kind of trace evidence to search for. Handling, packing and transporting material to the laboratory were regarded as vitally important. Although detectives were not expected to do laboratory analyses themselves, the report raised the question of their demonstrating some techniques as part of their training because they needed to know enough 'to give them the necessary insight into the possibilities of laboratory work in the various fields, so that they may the better understand the range of the subject'.[75]

Police forces required a certain amount of scientific apparatus, including photographic equipment and a dark room, appropriate space for examining and packing items of evidence, a binocular magnifier and an ultra-violet lamp which could be used to help examine items if ultra-violet facilities were not available nearby.[76] Cryptically, the report stated that other chemical apparatus might be required under the heading of 'police technique' but without specifying what that equipment might be. Nevertheless, the report did *not* recommend that detectives with some scientific training undertake some initial preparation of the evidence, emphasizing instead the difference between police technique and the actual examination of trace evidence, reinforcing a strict division of professional roles. The former was seen as an activity for the police, whereas the latter definitely fell under the purview of the expert in a laboratory.[77] Appropriate division of labour between scientist and detective was emerging.

The Detective Committee detailed exactly what laboratory analyses were required: ultra-violet and infra-red examinations; other chemical analyses; spectroscopic identification of particles, including metallic dust traces, paint and so on; facilities to examine documents and inks; comparison microscope analyses, particularly those involving tools and tool marks; weapons, bullets and cartridge cases; tests for bodily fluids, including semen and blood (for blood groups); microscopic analyses of hairs, fibres and other trace materials; photography; and pathological work.[78]

Just as division of labour between police and scientist was negotiated, so, too, was the question of whether a forensic science laboratory should be able to carry out all its own analyses or whether other laboratories, public analysts and other experts could be called upon. Under the older-style expert witnessing system, a police force employed an appropriate consultant for a particular case as required. Chief constables decided which expert was to be used, and lists of appropriate experts, including public analysts, had been drawn up and shared with other forces. The committee recognized that it would still be necessary to use independent experts for certain cases, but there were considerable drawbacks to relying completely on the independent expert system, and these the committee enumerated in some detail. It was recognized that much work would involve routine laboratory analyses for which an expensive external expert was not cost efficient; besides, the expertise of the new laboratories needed to be expanded if they were to be effective. If the committee's estimate of 10,000 cases or more in England and Wales requiring scientific analyses was realistic, this implied a quantity and range of work needed from forensic laboratories well beyond the capacity of a system of individual external experts.[79]

Cost was also an issue. Because it was a question of an individual fee, only the most serious cases or those thought likely to be winnable would be sent to an external expert; this had already suppressed the use of forensic scientific analyses. Of course, the new laboratories came with a cost but one that was to be shared between the Home Office and local forces, depending on size. If a police force was already paying what was effectively a fixed tax for laboratory services, they would use these services and this would encourage expansion of the laboratory system. Further reasons to move away from the individual expert system included the problem that such a system did not dovetail into police training on the application of science, it did nothing to encourage contact between police and scientists and it did not encourage the development of research in forensic laboratories.

In the committee's eyes such arguments pointed conclusively to the wisdom of establishing a network of forensic science laboratories.[80] This promised to be a virtuous circle; the more demand was created for laboratory services, the more would be the opportunity to train detectives to understand the potential of science as an aid to detection and the more would crime scene material be sent to the laboratories for analysis. The important thing was to establish the laboratories before demand escalated.[81] The proposed staffing of the new laboratories was envisioned to be appropriately modest. Recommended initial staffing was to consist of three to four scientists of whom one or two would be chemists, and the other posts would be for a physicist and a biologist, respectively. Expertise in photography and microscopy was essential. Despite these modest staffing requirements, this was at least a recognition that scientific specialisms were important in the new forensic science laboratories. Although close working relationships between police and the laboratories were desirable, it was vital that the laboratories operated as separate, autonomous bodies, with staff appointed by the Home Office, supplementing rather than superseding police work.[82]

The scientific staff appointed in the laboratories needed to have an appropriate level of authority, and this relates to some of the difficulties forensic scientists experienced in court appearances. Clearly, the person who undertook the scientific analysis had to be a demonstrably qualified expert in their field. This individual was, of course, a witness for the prosecution and had to have the ability to mount a good performance in court. The prosecution witness giving evidence had to be completely familiar with 'theoretical and practical aspects of the subject matter with which he is dealing and fully competent to meet cross-examination of the kind which may be brought against him if expert advice has been called in by the defence'.[83]

The Metropolitan Police Laboratory – uncertain beginnings

The setup and initial lack of success of the forensic science laboratory attached to the Metropolitan Police illustrated, at least in part, the difficulties of negotiating professional roles and carving out a space for the new breed of forensic scientists in relation to that occupied by forensic pathologists and independent forensic scientists on one side and the police service on the other. When the Metropolitan Police Laboratory opened its doors in Hendon in 1935, there must have been high hopes for its future. Its failure was down to a combination of management in the form of its first director, whose background was pathology rather than the natural sciences, an anti-scientific approach towards some of its work and the animosity that its association with the Police College to which it was attached engendered within the police.

The inauspicious way that forensic science was introduced into the Metropolitan Police and, in particular, the way that Cyril Cuthbert, appointed as police liaison officer on the opening of the laboratory, was treated, can be read as a harbinger of the early career of the new laboratory. It was not at all clear at the beginning whether the Metropolitan Police would welcome the introduction of scientific aids; indeed, the evidence suggests that they at best barely tolerated scientific analyses and even discouraged them. In Cyril Cuthbert's comprehensive account of the work of the Metropolitan Laboratory, *Science and the Detection of Crime*, he gave no hint as to his personal struggle to work in forensic science.[84] Fortunately, his colleague Hamish Walls had no such qualms about spilling the beans in a later paper on the history of the forensic science laboratories.

Before the setup of the lab, 'the Metropolitan police had their own private forensic scientist in the person of C.R. Cuthbert, a member of their C.I.D.'.[85] He had some medical and dental training but had not completed a degree, and he had taken evening classes in chemistry. Cuthbert taught himself all he could about the subject and brought together some equipment, including a '35-shilling second-hand microscope'.[86] He was occasionally consulted by his colleagues, and he established a network of contacts including forensic pathologist Bernard Spilsbury, analytical chemist Charles Ainsworth Mitchell and scientists from the Department of the Government Chemist.[87] He was just the kind of detective with a special aptitude for science that the Detective Committee hoped to identify in police forces. However, his superior officers initially regarded these activities as highly inappropriate and 'he was forbidden to waste police time and space on such nonsense'.[88] When Lord Trenchard, Commissioner of the Metropolitan Police, arrived on the scene, having heard of Cuthbert's work and realizing it could help form the basis of his new scientific laboratory, he asked to see the laboratory 'and was shown Cuthbert busy in the poky little cupboard . . . as a work place and wearing for the first time a white coat which had been hastily borrowed from a hospital'.[89]

The attitude towards Cuthbert's work shown by the 'Old Guard' was indicative of an anti-scientific attitude amongst senior officers in the Metropolitan force that did not bode well for the inauguration of the new laboratory.[90] The appointment of Dr James Davidson, a pathologist, who had been an assistant of Sydney Smith, Professor of Forensic Medicine at the University of Edinburgh, illustrates some of the professional issues at stake.[91] Lord Trenchard was pressing for a number of police reforms, including the proposed national medico-legal institute.[92] Walls recounted Trenchard arriving in the commissioner role 'full of grandiose ideas'.[93] The forensic science laboratory was, at least in part, designed to demonstrate the preeminence of the Metropolitan Police amongst British police forces.

So prestige, it seems, was certainly part of the story of the laboratory's origins, and this is reinforced by Trenchard's desire to persuade one of the well-known medico-legal experts to take the post of Director, even though he had been advised by Professor Tryhorn, a chemist, who was an expert in forensic matters and a member of the Detective Committee, to appoint a chemist. After consulting Bernard Spilsbury, Trenchard approached John Glaister Junior and Sydney Smith for suggestions for the director, possibly even hoping that one of these eminent professors could be lured down to London.[94] On Tryhorn's suggestion of a chemist, 'Glaister, whilst noting that he was not aware of the existence of Professor Tryhorn, thought the suggestion absurd'.[95] Glaister felt that an expert witness declaring himself to be a chemist would be off-putting to the jury (given that expert witnesses would have often appeared in court to give evidence) and that most cases would be medical. The Detective Committee was sensible to the fact that forensic laboratory expert witnesses would need to be able to perform confidently in court, but they judged that the vast majority of cases where forensic expertise was needed would be crimes against property, hence not requiring medical expertise. This was the complete opposite of Glaister's view. Although salary and conditions were not sufficiently attractive to lure a renowned medico-legal expert, unsurprisingly, both professors recommended that a medical man be appointed to the directorship of the laboratory. So Trenchard had to make do with Smith's former assistant, Dr James Davidson, whom he regarded as a stop-gap (see Figure 4.1).[96]

Appointed alongside Davidson were L. C. Nickolls, a chemist from the Department of the Government Chemist, who would later assume the role of director; Cyril Cuthbert, the police liaison officer, enthusiastic but largely self-taught, and a year later another chemist, C. G. Daubney and the physicist Hamish Walls. Walls got the job because the laboratory commissioned a spectrograph as its first piece of instrumentation and he had experience of such instruments as part of his PhD.[97] Walls was still in his twenties when new opportunities for full-time forensic

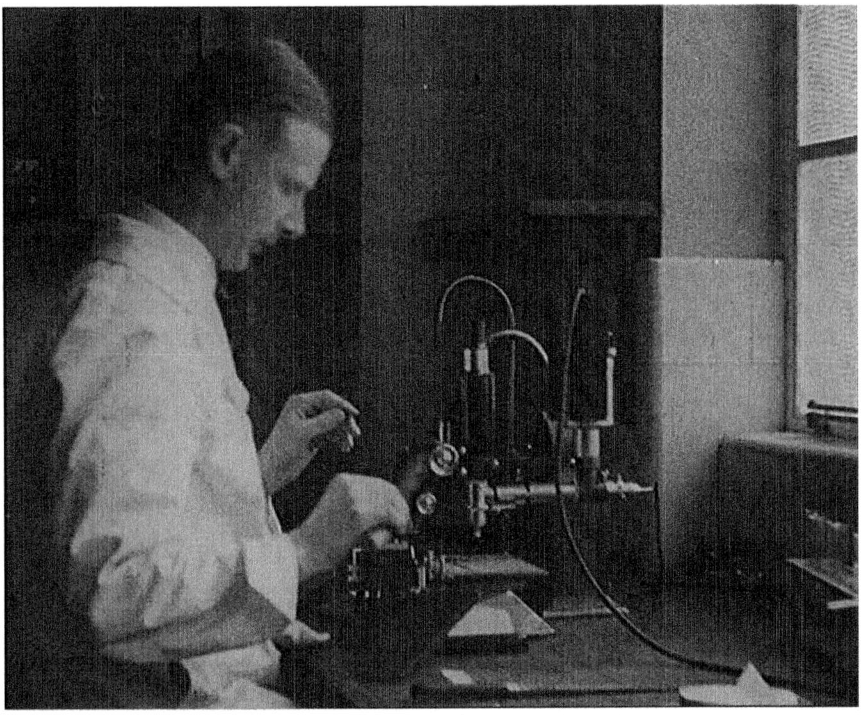

Figure 4.1 Dr James Davidson at the opening of the Metropolitan Police Laboratory[98]

scientists were opening up in England and Wales. With only a couple of years' post-doctoral university research experience and a year with the chemical company, ICI, he was totally inexperienced in forensic matters and would hardly have had substantial professional connections at that stage. Indeed, if his autobiography is to be believed, he was appointed to the role, having made enquiries of the Metropolitan Laboratory Director, James Davidson, whom his family knew slightly as both came from Edinburgh. The laboratory needed a physicist to operate its new spectrograph and here he was. Professional delineations between scientific disciplines might be evident, but there was clearly some flexibility in the early years. As Walls described, since analytical chemistry focused on test tubes and solutions, a spectrograph was not regarded as quite proper for a chemist to use.[99] Although his background was in physical chemistry, as the operator of this instrument, he was now a physicist, rather, a 'physicist' because as he put it, in his characteristically picturesque style, 'I was about as much a real physicist as a bookmaker's tout is a yogi'.[100] Walls was to move to the Bristol Forensic Science Laboratory in 1946, becoming Director of the Newcastle Laboratory in 1958, before returning to the Metropolitan Laboratory as Director in 1964 and retiring in 1968.[101]

Davidson was not a great success as Director, and the laboratory did not flourish under his stewardship, although this may have been related more to his personal qualities rather than specifically the fact that his background was medical.[102] Yet there were a number of perceived issues relating to the potentially medical bias of the new laboratory. Trenchard had been keen to set up a medico-legal institute and this was not it. Dixon's bias was always more towards the scientific, given that the vast majority of chargeable offences were crimes against property which demanded scientific rather than medical analysis.[103] This view was borne out by senior forensic scientists. C. T. Symons, the first Home Office Forensic Science Adviser, formerly Government Analyst in Ceylon, and therefore well versed in forensic chemical and other scientific analyses, pointed out: 'doctors were not fully aware of the scientific importance of some of the material they handled'.[104] The new Commissioner, Sir Philip Game, who succeeded Trenchard at the end of 1935, favoured the older system of appointing independent experts rather than using the services of the laboratory and this also tended to undermine its professional position.[105]

Added to these difficulties, the lab's association with the Police College sited alongside at Hendon added to its unpopularity amongst the ranks. The 'Old Guard' at Scotland Yard treated it with suspicion and little work was passed to it.[106] Echoing the Detective Committee's view on educating the police as to the use of forensic science, Walls pointed out in his autobiography that the laboratory was a new idea and the police did not know how to use it. At the time the laboratory seemed to take the attitude: '"Here we are. Take us or leave us." In consequence we were more left than taken.'[107] Without the support of senior Metropolitan Police circles, and with a disgracefully low case load, the laboratory was in serious danger of closing in 1938.[108]

An education campaign was needed, but this did not happen until after the war. A British Pathé film from 1946, *Science Fights Crime*, brought to the public's and

Figure 4.2 Dr Hamish Walls using a comparison microscope at the Metropolitan Police Laboratory[109]

police's attention the detective training courses which had now resumed, with training on handling scenes of crime, particularly murder scenes, with carefully distributed trace evidence and a crime scene dummy: 'Mary, the most murdered model in Hendon'. Alongside shots of fingerprinting techniques, a comparison microscope for comparing bullets and ultra-violet photography (see Figures 4.2 and 4.3), Detective-Superintendent Rundle intoned: 'Training and good perception backed by modern science puts us one ahead of the criminal.'[110]

The regional forensic science laboratories

The Metropolitan Police Laboratory was the flagship for the new forensic science, so its success or failure was highly visible. The regional laboratories generally fared better; however, it would be wrong to suggest that the initiation of the regional laboratories was without its own set of difficulties to overcome, not least because it represented the introduction of a state system into what had been up until then a free market.[111] The Nottingham Laboratory was already operating in the 1930s as a police laboratory, its success considerably aided by the enthusiasm of Chief Constable Popkess, who was a keen supporter of science in the service of

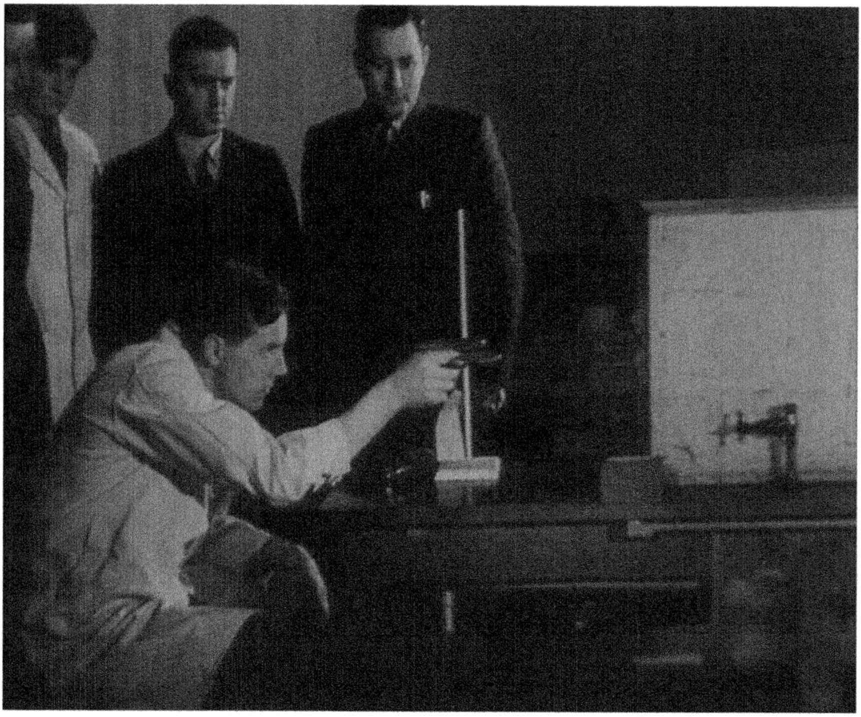

Figure 4.3 A ballistics experiment at the Metropolitan Police Laboratory

crime detection. He did not want to wait for the Home Office scheme to be fully operational but pressed ahead with his own plans for a laboratory.[112] Popkess contributed a number of articles on 'scientific aids' to the pages of the *Police Journal*. The first scientists working at the Nottingham Laboratory were appointed as police constables, thus raising questions of impartiality, a matter of some concern to C. T. Symons in his role as Home Office Forensic Science Adviser. Nevertheless, Dixon was keen that Nottingham should be designated the first Home Office Regional Laboratory as it was working efficiently because Popkess was such an enthusiastic champion of scientific aids and because Nottingham was in a significant crime area. Dixon wanted the laboratory to be separated from police control and the costs to be spread amongst the police forces of the region. Running in its intermediate stage from 1936, the Nottingham Laboratory was regionalized to become the first Home Office Regional Forensic Science Laboratory from 1938.[113]

Part of the reason for the Nottingham success was the willingness of its chief constable to reconfigure his existing police laboratory according to the Home Office scheme. However, it was perfectly reasonable for a chief constable to be as enthusiastic about scientific aids in detection as Popkess was, but to be less than enthusiastic about the Home Office model of laboratory organization. One such chief constable was Anley from Derbyshire who had his own police laboratory staffed by police officers, and who did not wish to be tied to using the services of a regional laboratory exclusively. He preferred to call in outside experts for scientific analyses beyond the capabilities of his laboratory.[114] The strength of Anley's view can be seen in his minuted reservation to the Detective Committee's report, the only reservation listed from any of the committee's members. He assented to the overall report with the exception of proposed regional clearing houses and the requirement for chief constables to send evidence for scientific examination to any one laboratory on the basis that, as chief constables were responsible for crime detection in their force, they should be able to consult whichever experts they wished.[115]

The 'Derbyshire' view was reinforced by the publication of a book on scientific aids by two senior officers of the Derbyshire force, which advocated the use of individual experts for forensic evidence.[116] Although this book endorsed the 'scientific aids' approach and was published by the *Police Journal*, it took a different line from the mainstream of the 'scientific aids' movement. Although much of their book echoes the kind of advice on trace evidence given elsewhere – for instance, it dealt with packing alongside topics such as bloodstained articles and fibres found at the crime scene – it is notable for suggesting individual names and contact details of experts who could be contacted to advise on particular kinds of evidence, e.g., Roche Lynch for human flesh, Ainsworth Mitchell for fraudulent documents.[117]

Regional laboratories at Cardiff, Bristol, Birmingham and Preston were to follow.[118] Dixon was keen to develop a laboratory in Birmingham, as he saw it as a probable location for his central laboratory. Additionally, he wished to include expertise in pathology at Birmingham, partly because he understood the importance of integration of forensic science analyses with medical analyses and also because he was following the principle of spreading specialist expertise across the regional laboratories. As Ambage suggests, he may also have been keeping a weather eye on the situation regarding a medico-legal institute as recommended by the Scientific Committee.[119] James Webster, a police surgeon with a wide-ranging background in police work, was appointed to head the Birmingham Laboratory in 1937 and to act as a second Forensic Science Adviser to complement the expertise of Symons who was a chemist. Holden, Director of the Nottingham Laboratory, took over Symons's role unofficially after the latter's death in 1937.

The situation with the North Western Laboratory at Preston was somewhat different. There were some questions as to what should be the appropriate experience of the director of the proposed laboratory. The idea of a pathologist heading the laboratory was raised, although opposed by Holden and Webster. There was even the suggestion that Sydney Smith should advise, of which suggestion Holden remarked: 'I imagine that he is somewhat imbued with the idea that only medical men are really competent to deal with forensic work.'[120] In the event Smith was not asked to advise, and James Brierley Firth, a chemist, with a wide-ranging university and industrial background, was appointed to head up the new laboratory.[121]

Firth's experience was quite different from that of the Director of the Metropolitan Laboratory, James Davidson. By the time of his Home Office appointment,

Firth was around fifty years of age, with a successful industrial and university career under his belt. Whilst lecturing at University College, Nottingham, he became a consultant for the Ministry of Agriculture and Fisheries surveying the River Trent after the River Pollution Act was passed. This led him to become a consultant on water supplies, sewage and effluent to several local authorities.[122] These roles involved many appearances as a scientific witness in a variety of courts and at Ministry of Health inquiries. He had already spent hours in the witness box.[123]

In addition to his court work as a scientific expert witness, he was asked to help Nottingham Police on several occasions when they needed a chemist, and this started him on the path of lecturing to various police forces on the use of science in crime detection.[124] It was no doubt this combined experience of using science to detect crime and, very importantly, his skill as a scientific expert witness in the court which drew him to the attention of Arthur Dixon when he was looking for experienced scientists to lead the new forensic science laboratories. Firth, no doubt, appeared to be a safe and very experienced pair of hands.

Although a newspaper article reporting Arthur Dixon's speech on the laboratory's opening in 1939 hints at a grand opening of a shiny modern laboratory, the reality was that the lab consisted of two rooms in a terraced house with a staff of two, Firth and a constable with a degree.[125] Although they had the cooperation of the Lancashire police forces, some forces treated them with suspicion at first.[126] It was no doubt Firth's considerable experience in working with and training police that helped to break down barriers and establish the cooperation and teamwork between police and laboratory which he knew was vital to the lab's success. Having been involved in the area for years, Firth continued police instruction in scientific matters in his retirement and believed that such instruction encouraged a scientific approach, instructed in methods of taking and preserving trace evidence and informed police officers about the kinds of things the lab could do.[127]

The official openings of forensic science laboratories were reported in national and local newspapers; views expressed in such reports were generally optimistic about the role of forensic science in the fight against crime. Undoubtedly, official laboratory openings afforded splendid public relations opportunities where regional and national newspapers could report the positive messages of senior civil servants, local government officials and chief constables reiterating the importance of the police–scientist relationship in solving existing crimes and deterring new ones. For instance, the *Manchester Guardian* reported the official opening of the North Western Laboratory in Preston in 1939 by detailing the lab's impressive list of equipment and facilities, which appeared to have grown somewhat from the initial two rooms in a terraced house, and by reporting Dixon's opening speech.[128] Dixon stated that he regarded the North Western Laboratory as a further milestone in the development of the subject. He emphasized close collaboration between police and scientist and the necessity of avoiding the situation in other countries where scientists immersed themselves in their own academic work without contributing sufficiently to police work. The laboratory undertook a range of analyses covering ordinary forensic requirements, and it contacted other forensic science laboratories if support was needed. This meant that the laboratory did not have to cater for all possibilities: the power of the system lay in the combined strength of the laboratories. Perhaps mindful of possible insurrections from Derbyshire, and also question marks over the Metropolitan Laboratory's association with the Hendon Police College, Dixon stressed the North Western Laboratory's independence: the Director was concerned solely with scientific analyses.

In the same *Manchester Guardian* article, the approbation of Captain Hordern, Chief Constable of Lancashire, was recorded. In the first half of 1939, Hordern noted that the lab dealt with three murder cases, thirty-two counterfeit coin cases, thirty-three offences against the person (chemical and biological analyses) and fifty-six property offences, the latter including break-ins, larceny, false pretences, arson, malicious damage and identification of bodies where Firth's analysis had been especially helpful.

> His department has been able to give clues as to the perpetrators of crimes when, to our point of view, there appeared to be no clues, and we have been given a lead on many occasions where we had no lead before.[129]

John Maxwell, Manchester's Chief Constable, voiced the opinion that the cooperation between police and scientists that such laboratories fostered encouraged police officers to search for ever smaller details so the problems of the past where much material must have been overlooked would now be avoided.[130]

Contrast the public relations exercise on behalf of forensic science at the official opening of the North Western Laboratory reported above by the *Manchester Guardian* in 1939 with the argument for the proposed medico-legal institute which formed the basis of a newspaper report of the opening of the Metropolitan Police Laboratory in the *Daily Mail* in 1935.[131] It is hard to believe that both articles were reporting the opening of forensic science laboratories. The Metropolitan Lab opening was headlined 'Lord Atkin's Plan', 'Resources to Aid Whole Nation' and 'Police Laboratory Opened' but was largely given over to Lord Atkin's suggestion for a medico-legal institute, the importance of appointing a chair in forensic medicine at the University of London as its head and instituting a diploma for coroners.[132] The Home Secretary, Sir John Gilmour, 'nodded approval' to these plans, mentioning that he had set up a committee to advise on the new laboratory. The last third of the article advised that there were two purposes of

> . . . the Police College's new laboratory, which includes all the apparatus of detection familiar to lovers of Sherlock Holmes . . . curriculum of selected police officers . . . what to look for at the scene of a crime . . . a research department where examination of specimens . . . in the detection actual crime can be scientifically conducted.[133]

It is important not to read too much between the lines in a short newspaper article. Even so, a number of points stand out in relation to this article which gave the impression that at the opening ceremony, considerable attention was given to arguments for a medico-legal institute (something that the lab was not). The journalist described the laboratory as the 'Police College's laboratory', although it was the Metropolitan Police's laboratory, and did not strictly belong to the college. This tended to diffuse the autonomy of the laboratory and reinforced the suggestion that the lab was seen as associated with the college right from its inception, to its detriment, as Walls suggested.[134] In any case, it would have been hard for the laboratory to separate itself from the college, given that it was co-located and so heavily involved in the training of detectives.

Chapter 6 notes that a nod to Sherlock Holmes was popular amongst forensic scientists' writings. Nevertheless, invoking the fictional detective can have the effect of lessening the seriousness of a point, especially when one considers that Sherlock Holmes's laboratory facilities were distinctly amateur. However, this was mitigated by a reference to fictional medico-legal specialist Thorndyke![135] Finally, the journalist talked of the two functions of the laboratory. Undoubtedly, one of the functions was police education in crime scene management, but the second function, namely scientific examination of crime scene specimens, was described as 'research', which has the tendency to make such activities seem somewhat removed from the day-to-day business of analysing trace evidence. It is important not to read too much into a newspaper article, intended to be a short, ephemeral report; nevertheless it certainly conveyed a different impression from the more focused message delivered at the opening of the North Western Laboratory.

A British Pathé newsreel of the opening of the laboratory showed a set of VIPs on a platform looking as stiff as their starched wing collars.[136] Indeed the platform party (about twenty officials) considerably outnumbered the staff of the laboratory, who were initially six.[137] In his opening speech, Sir John Gilmour referred to the opening of a new chapter in scientific research, and the next scene panned to a laboratory with about six microscopes, each with its separate lamp and power supply for the lamp, spread over three lab benches with traditional laboratory Bunsen burner–type gas sources and sinks. This was the training laboratory which would be used to help train detectives. Although a comparison microscope for the comparison of bullets was presented later in the film, the training laboratory was portrayed with a somewhat empty feel, not helped by the way that no laboratory or detective staff were filmed at work at the main benches. Despite this, the fact that the opening of the laboratory was widely reported and recorded in a British Pathé newsreel were indications of its perceived significance.[138]

The thorny question of whether the defence should have access to the services of the forensic science laboratories was an issue.[139] Dixon's view was that the defence should be informed if an analysis favoured their case; however, defence lawyers could not explicitly consult Home Office experts. His speech at the opening of the North Western Laboratory said as much, to the extent that he claimed that if the evidence favoured the defence, then he believed that they should have the results even if it threatened to 'blow up' a police case.[140] Apart from the fact that the laboratories were set up explicitly as a service to the police, the spectre of the undesirable situation where experts were pitted against one another all too readily raised its head. Scientists battling in court could have undermined the epistemological status of the emerging discipline of forensic science and could also have had a deterrent effect on recalcitrant police forces, who would be less likely to use their regional laboratory if they thought their evidence would be contradicted by an expert from another laboratory, either independent or under the aegis of the Home Office.[141] Although many independent expert witnesses remained, the new state laboratories were in effect taking control of science in the service of the prosecution. In constructing the nascent forensic science laboratory network, Dixon was keen to appoint good all-rounders – male, of course – with excellent qualifications and experience dealing with scientific evidence for police or legal purposes.[142] At least part of this recruitment strategy was based on the knowledge that such experts would need to be able to mount a creditable performance in the witness box, as we have already seen.

The 'scientific aids' movement

Although there were clear differences in approach between the Metropolitan and regional laboratories and amongst regional chief constables, nevertheless the 1930s saw a major push for a division of labour between police officer and detective. This went hand in hand with the concomitant professionalization of the detective role, much discussion as to the delineation of proper activities for police officer, detective and scientific laboratory, in relation to the use of science in crime detection. This included the development of a portfolio of procedures for officers to undertake at the crime scene and the development of forensic science. This can be characterized as the 'scientific aids' movement, the British equivalent of *criminalistics*. Those involved in promulgating the use of science in crime investigation included members of the Detective Committee, including Arthur Dixon as the most senior figure, senior police officers and a number of scientists who advocated a scientific approach to crime. From the 1930s onwards, they were anxious to push for scientific aids in criminal investigation and for the whole administrative and bureaucratic edifice necessary to make the new arrangements work, although there was yet to be complete agreement about what the institutional and administrative arrangements should be. This movement was centred on policing, but it included scientists leading the setup of the new forensic science laboratory network and developing training courses, such as C. T. Symons and F. G. Tryhorn, one of the leading scientists in the movement, who would eventually become Director of the Harrogate Laboratory and then Home Office Forensic Science Adviser.[143]

Scientific aids were not science. They could more properly be portrayed as a part of criminalistics, or forensic technoscience, as the story includes the wrap-around of administrative and technical apparatus. Scientific aids were promoted to achieve the delineation and understanding that police officers and detectives had to have as to the possibilities of science in the detection of crime. It was part of the process of achieving 'administrative objectivity' for the procedures that attended the development of forensic science. Considering more recent times, Lynch argues that

DNA profiling in forensic applications has been granted a high level of epistemic certainty because the processes surrounding it have achieved acceptance within criminal justice authorities as administratively objective.[144] 'The "administrative objectivity" of DNA evidence rests upon observable and reportable bureaucratic rules, records, recording devices, protocols, and architectural arrangements.'[145] Despite the tweaks and workarounds always needed to maintain administrative procedures, the accepted certainty of forensic DNA profiling has permitted administrative closure to the extent that these protocols, if shown to be properly administered against a set of agreed-upon processes, will not be challenged in court.[146] It is the business of working out the rules and protocols surrounding scientific analyses that gives them demonstrable legal credibility in the courtroom, to achieve administrative closure as far as possible, and working out the rules and procedures also requires working out and getting agreement on which activity is undertaken by which individual in a designated role. By the 1930s, fingerprinting had achieved administrative closure, and this was partly why it could be relegated to the realm of police technique. Just as with fingerprinting early in the twentieth century and with modern forensic DNA profiling, 1930s forensic science required the achievement of agreed-upon administrative objectivity, and the 'scientific aids' approach was a considerable force in this achievement.

It was the fact that publications surrounding scientific aids were closely connected with the police, the new laboratory system and the Detective Committee and its mission which marked them out as embodying this distinctive approach. The history of 'scientific aids' is to be found in the Home Office Scientific Aids (1936) instructional pamphlets; the more specialized, less widely circulated and shorter-lived Home Office Forensic Science Circulars (from 1936); the Detective Committee report (1938); a book on the detection of crime published by the *Police Journal*; a few articles appearing in the short-lived *Metropolitan Police College Journal* and a substantial series of articles, particularly those written by Tryhorn, in the *Police Journal* itself.[147] It is tempting to include in this group detailed descriptions of forensic laboratory life, such as Nickolls's *The Scientific Investigation of Crime* or Cuthbert's *Science and the Detection of Crime*.[148] Valuable as these are for describing the setup of the forensic laboratory system, and they are certainly in tune with the message of the material produced in the 1930s, they were written after this period and therefore were not part of the story of working out how to do things in the 1930s. There were a number of popular and textbook works in the penumbra of 'scientific aids'; these contributed to arguments for the development of forensic science, and their message was generally in tune with the 'scientific aids' authors, yet these tended to be slightly more peripheral to the 'scientific aids' mission as it played out in the UK. Alfred Lucas's textbook *Forensic Chemistry*, for no other reason that it was so widely quoted, can be included, as can the popular works of independent forensic scientists such as Charles Ainsworth Mitchell (both authors were writing rather before the 1930s) and independent forensic scientist Julius Grant, whose semi-popular work *Science for the Prosecution* (1941) had much to say about the institutional setup of forensic science in England alongside traditional scientific details.[149]

The *Scientific Aids to Criminal Investigation* instructional pamphlet (1936) for police officers provided detailed instructions on what to do at the scene of a crime, including how to judge the value of scientific evidence (where the message was that it was better to send too much material than too little), how to search for material and examine the scene of crime and very detailed instructions as to packing, handling and labelling. There was no science in the *Scientific Aids to Criminal Investigation* pamphlet. Rather it was a manifesto for British criminalistics, in handy pocket guide form, with highly detailed explanations of how the crime scene was to be managed by the police officer. The introduction proselytized the 'scientific aids' message, namely that 'the scientist can often render valuable assistance to the police if he is afforded an opportunity to examine articles connected with the scene of a crime'.[150] In cases where no witnesses were available or where witnesses' evidence was in conflict, scientific evidence could provide valuable circumstantial evidence to the extent of acting as a kind of silent witness, the traditional view of the accepted objective nature of physical evidence.

> Scientific evidence, when properly interpreted, may thus provide, in certain classes of cases, a kind of proof which never lies and never alters its tale, and which, if placed before the Court by a competent witness, can be seen by Judge and jury for themselves.[151]

A second edition of the pamphlet was published in 1950 with the addition of a number of block and link diagrams of a 'scientific management' sort and a brief additional section on type and handwritten documents. This edition was reprinted in 1964. Apart from some phrases printed in bold typeface, the introductions were identical in all three editions. Indeed, it is striking how few amendments were evident between the different editions, implying that the *Scientific Aids to Criminal Investigation* pamphlet enjoyed a shelf life of almost thirty years substantially unaltered.

The publication of a new journal of policing from 1928: *Police Journal: A Quarterly Review for the Police Forces of the Empire* heralded the new era of scientific aids in police detection, and, as its name suggests, this was to be seen as not only a matter for the UK. Policing in the Empire was an integral part of the story of the introduction of science to crime detection. The journal's first editorial argued that a link between the police forces of the Empire was lacking. Rather than emulating the local publications available to police forces, this was to be regarded as a service journal 'in which the views of the professional police officer and the scientific authority on criminal matters can be made available at once to police officers throughout the Empire'.[152] The new journal was to be 'devoted to the work, organization, history and development, scientific, legal and sociological, of the police forces of the British Empire.'[153] Although the journal was officially independent, it received the patronage (although it is not possible to say how much that amounted to financially) of the Home Office, Scottish Office, Dominions Office, Colonial Office and India Office, so it was set up with the backing of the offices of state. In the first year it covered policing matters in the Empire and beyond, including the

UK, Italy, Palestine, Iraq, India, China, Canada, Ceylon, Denmark, South Africa, Burma, Belgium, France and East Africa. The journal was intended to cover topics 'administrative, technical, historical, personal, medico-legal or sociological' so long as it was of interest to police officers. 'The scope of police work is steadily widening; and it is becoming in an increasing degree scientific.'[154] It is no surprise that the new journal emphasized policing the Empire and that it should connect scientific policing with imperial policing, given the colonial origins of a number of scientific approaches towards criminal identification and crime scene management, as the last two chapters have described.

The *Police Journal* is remarkable for its high level of scientific content. It matched and even overtook the longer-running *Medico-Legal and Criminological Review*, the journal of the Medico-Legal Society, as the major UK periodical for publishing innovations or proposed innovations in forensic science in the late 1920s and 1930s. In its early years, of typically fifteen articles per issue, three or four were specifically on scientific matters, and a number of prominent forensic scientists, medico-legal experts and senior police officers wrote on forensic innovations in its pages. Importantly it was at the heart of the 'scientific aids' movement; it was, to some extent, the house journal of the 'scientific aids' approach. In the first volume Locard's article on the importance of dust was published, forensic chemists Rhodes and Ainsworth Mitchell wrote about scientific evidence in the detection of crime and Glaister and Sydney Smith, the latter a regular contributor to the journal throughout the 1930s, wrote about bloodstains and hairs and ballistics, respectively.[155] Willcox, as Home Office Analyst, produced an article on toxicology.[156] Although subsequent issues could not always match such a lineup of forensic star performers, nevertheless right from the beginning an array of senior figures in the world of forensics was represented in the pages of the *Police Journal*.

A few years later, in 1933, the *Police Journal* was to reposition its mission. Still focusing on policing at home and in the Empire, Dixon's introduction to volume six announced an emphasis on real cases and practical aspects of police technique and equipment and the didactic role which the journal should have, particularly for the forces of England and Wales who were organized as local forces increasingly dealing with national problems, which was one of Dixon's perennial messages.[157] Unsurprisingly, given his interest in science and, in particular, forensic sciences, Dixon indicated the topics and applications that the *Police Journal* was to concern itself with:

> motor transport, wireless telegraphy and telephony, and other means of rapid communication . . . the more effective organization of records . . . the application of scientific aids to detective work, with which should go the better instruction of all ranks, uniform and detective, regarding the possibilities available to the police in this direction.[158]

This was a shopping list defining the technoscientific aspects of scientific aids. It was, of course, a list of the topics falling under the remit of the Detective

Committee; hence the *Police Journal* was therefore an effective way of discussing and disseminating the work and ethos of the Detective Committee to British and colonial police forces. The *Police Journal* nailed its 'scientific aids' banner firmly to the mast of modern policing in publishing Tryhorn's series of 'scientific aids' articles on a wide range of forensic topics between 1935 and 1938: packing, marks and impressions, blood and bloodstains, fires and glass.[159]

Tryhorn's 'scientific aids' articles were a good basis for demonstrating the range of interests of the movement. Supreme amongst these was the issue of packing. He opened his series of articles on scientific aids in the *Police Journal* with an article titled 'The packing of exhibits'.[160] Packing seems like a quotidian subject hardly falling within the remit of forensic science. Yet it was the vital plank on which all the scientific analyses in the forensic laboratory rested. We would probably be somewhat surprised if a twenty-first century professor of chemistry concerned himself or herself with the minutiae of packaging, the exact dimensions of boxes and detailed lists of the contents of such boxes in print. But Tryhorn was no ordinary professor of chemistry: as a forensic scientist not quite so *manqué*, and a forensic scientist who was at the forefront of the discipline as it emerged, he knew that the details of crime scene management were just as important as the scientific analyses. Indeed, he knew that the scientific analyses were useless without properly managed evidence from a crime scene. Today we would call this management of trace evidence to preserve the chain of custody; then it was called 'packing'.

Tryhorn was by no means the first to emphasize the virtues of attending to the detail of packing. A number of forensic experts were saying much the same thing in the first half of the twentieth century. Gross's translators described packing in relation to blood-stained materials in his *Criminal Investigation*, although it is interesting that this work did not say as much about this as one might expect, given Gross's emphasis on the management of crime scenes.[161]

Indeed at the beginning of his article on packing, Tryhorn noted that several authors working on scientific aids in criminal investigation had written on such matters, specifically citing Lucas's, *Forensic Chemistry and Scientific Criminal Investigation* for listing common problems with packing and the types of damage that evidence could sustain while it was being moved.[162] Indeed the reference to Lucas was a reminder that it was not quite so unusual for a chemist to be concerned with packaging crime scene materials after all. Although Tryhorn did not specifically allude to this, as an expert analytical chemist, Lucas explained the reason why the chemist was an appropriate type of scientist to deal with the minutely specified details of packing requirements. This was, as Lucas argued, all to do with the character of chemists and their training in working with the details in chemical analyses, often working with very small amounts of substances which required to be analysed.

Lucas, as an analytical chemist and an archaeological conservationist, was an important source of knowledge on packing forensically important materials, as both roles involved consideration of careful ways of preserving fragile material; it is no exaggeration to say that Lucas was one of the world's experts in packing.

Indeed, he was only too aware of the results of bad packing, of not ensuring that items were sufficiently secure, using thin cardboard boxes so items would break or leak out of containers.[163] It is, therefore, unsurprising that Tryhorn should draw on Lucas's seminal work.

Although the virtues of proper packing of exhibits could hardly be disputed by anyone dealing with forensic evidence, having the proper material available for packing was another matter. Some of the Detective Committee's list of required equipment, even the simplest packing materials, would remain something of a pipe dream for some time. There is evidence to suggest that there was a considerable shortage of materials and equipment in the Metropolitan Police Laboratory in its early years at least. Resources may have been sparse in the 1930s, but things were often even worse after the war with rationing and other post-war shortages. John McCafferty, appointed as Cyril Cuthbert's assistant after the war, was fairly shocked at the poor facilities on offer: 'an almost complete lack of cash for proper facilities available to the scientist or policeman for proper scientific work'.[164] Shortages of materials meant that there were severe difficulties in following even the simplest instructions to bring in materials from the crime scene for inspection by the lab.

> In the early days absolutely no facilities were available to officers for packing exhibits and they had to improvise by begging from the station stationery store or buying out of their own pockets such things as wrapping paper, small boxes or tins, or reusing materials such as old envelopes and sometimes even newspaper.[165]

Indeed envelopes, even when they were available, often ruined the forensic evidence in a case. Take the question of safe breaking as an example. In the post-war period safes were packed with a characteristic fire-proof powder.[166] So if a man was suspected of blowing open a safe, his clothing would be taken for examination along with a control sample of the powder as a comparison. However, if this sample were sealed in an envelope it might easily leak from the envelope corners onto the clothing, thus ruining the whole case. It was hard enough to get police officers to appreciate these dangers. 'But it was even harder to get the purse strings opened even just a little to enable suitable containers and packaging to be made available so these dangers could be obviated.'[167]

Conclusion

The professionalization of the detective role, with the push towards incorporating a scientific approach to detection to cope with a rising tide of crimes against property, were major elements contributing to the development of forensic science laboratories and scientific aids in policing. Undoubtedly, Arthur Dixon was a major figure and the work of the Detective Committee, in particular Sub-committee D, did much to promote the necessary changes through spelling out requirements for detectives' training, how scientific aids were to be incorporated into detection and the need for a network of forensic science laboratories. The setup of the initially less-than-successful Metropolitan Police Laboratory was soon followed by a number of regional forensic science laboratories. The arguments for 'scientific aids' made in detail, mainly through the pages of the *Police Journal*, but also through the *Scientific Aids* pamphlets, was a way of delineating the role of the detective or investigating officer and indicating what was to be done at the scene of crime and educating the police in the possibilities of science in the detection of crime. Despite uncertain beginnings, by the outbreak of war, forensic science laboratories were in operation in England and Wales and a discipline called forensic science carried out by forensic scientists was in existence. The next chapter considers the careers of those forensic scientists, both those inside and outside the Home Office system.

Notes

1 P. Knepper, *The Invention of International Crime: A Global Issue in the Making, 1881–1914*, Basingstoke and New York: Palgrave Macmillan, 2010, p. 33.
2 S. A. Cole, *Suspect Identities: A History of Fingerprinting and Criminal Identification*, Cambridge, MA: Harvard University Press, 2001, pp. 58–59.
3 J. Tyndall, *Scientific Use of the Imagination: And Other Essays*, London: Longmans, Green, and Co., 1872. Quoted by Sherlock Holmes in A. Conan Doyle, *The Hound of the Baskervilles*, London: George Newnes, 1902, available at https://www.gutenberg.org/files/2852/2852-h/2852-h.htm, accessed 30 April 2015.
4 Troup Committee, *Identification of Habitual Criminals*, London: HMSO, 1894.
5 Belper Committee, *Method of Identification of Criminals by Measurements and Finger Prints*, London: Printed for Her Majesty's Stationery Office by Wyman and Sons Ltd., 1901.
6 A. M. Joseph, 'Anthropometry, the police expert, and the Deptford murders: The contested introduction of fingerprinting for the identification of criminals in late Victorian and Edwardian Britain', in J. Caplan and J. Torpey (eds), *Documenting Individual Identity: The Development of State Practices in the Modern World*, Princeton and Oxford: Princeton University Press, 2001, 164–183, p. 168.
7 Ibid., p. 169.
8 Ibid., p. 172.
9 Ibid., p. 173.
10 HMSO, *Home Office: Report of the Departmental Committee on Detective Work and Procedure*, London: His Majesty's Stationery Office, 1938. The role of the Detective Committee is described later in this chapter.
11 Joseph, 'Anthropometry', p. 174.
12 HMSO, *Detective Committee*, Vol. 5, p. 8.
13 F. G. Tryhorn, 'The packing of exhibits', *The Police Journal*, 1935, 8 (1): 19–26; A. Lucas, *Forensic Chemistry*, London: Edward Arnold, 1921, especially Chapter one 'Introduction', pp. 1–15.
14 Joseph, 'Anthropometry', p. 167.
15 See C.R.M. Cuthbert, *Science and the Detection of Crime*, London: Hutchinson, 1958, for Cuthbert's description of the work of the Metropolitan Police Laboratory.
16 E.g., Tryhorn, 'The packing of exhibits'.

17 For books on scientific aids, see W. M. Else, and J. M. Garrow, *The Detection of Crime: An Introduction to Some Methods of Scientific Aid in Criminal Investigation*, London: The Police Journal, 1934; R. Morrish, *The Police and Crime-Detection To-Day*, London: Oxford University Press, 1946.

18 J. B. Firth, *A Scientist Turns to Crime*, London: William Kimber, 1960.

19 P. Knepper, *The Invention of International Crime: A Global Issue in the Making, 1881–1914*, Basingstoke and New York: Palgrave Macmillan, 2010, p. 36.

20 Ibid., p. 37.

21 A. Dunlap, 'Science versus practical common sense in crime detection', *American Journal of Police Science*, 1931, 2 (4): 322–327.

22 Ibid., p. 322.

23 Ibid.

24 Ibid.

25 H. J. Walls, *Expert Witness: My Thirty Years in Forensic Science*, London: John Long, 1972.

26 For instance, such concerns were then not so very different from the widespread fears that computers would take people's jobs, or at least alter them in unacceptable ways, views which were prevalent in relatively recent times, especially in the 1980s See U. Huws, *Your Job in the Eighties: Woman's Guide to New Technology*, London: Pluto, 1982.

27 Knepper, *The Invention of International Crime*, p. 35.

28 T. A. Critchley, *A History of Police in England and Wales 900–1966*, London: Constable, 1967, p. 176.

29 K. Laybourn, and D. Taylor, *Policing in England and Wales, 1918–39: The Fed, Flying Squads and Forensics*, Basingstoke and New York: Palgrave Macmillan, 2011, pp. 5–7.

30 J. Maxwell, 'The English police system: General developments and outstanding features', in L. Radzinowicz and J.W.C. Turner (eds), *Penal Reform in England*, London: Macmillan, 1946, 60–97.

31 Laybourn and Taylor, *Policing in England and Wales*, p. 11.

32 M. Foucault, *Discipline and Punish: The Birth of the Prison* (trans. A. Sheridan), London: Allen Lane, 1977.

33 J. Maxwell, 'The English police system: General developments and outstanding features', in L. Radzinowicz and J.W.C. Turner (eds), *Penal Reform in England*, London: Macmillan, 1946, 60–97, pp. 82–83.

34 P. Rawlings, *Policing: A Short History*, Abingdon and New York: Routledge, 2014 (first published 2002 by Willan Publishing), p. 185.

35 Ibid., p. 186.

36 See Laybourn and Taylor, *Policing in England and Wales*, for a notable exception.

37 H. Shpayer-Makov, *The Ascent of the Detective: Police Sleuths in Victorian and Edwardian England*, Oxford and New York: Oxford University Press, 2011, p. 46.

38 Ibid., p. 63.

39 Ibid., p. 64.

40 N. Ambage and M. Clark, 'Unbuilt Bloomsbury: Medico-legal institutes and forensic science laboratories in England between the wars', in M. Clark and C. Crawford (eds), *Legal Medicine in History*, Cambridge: Cambridge University Press, 1994, 293–313, p. 293.

41 Laybourn and Taylor, *Policing in England and Wales*, p. 82. Detective posts amounted to less than 5 per cent of the total police force in England and Wales in 1928–1929, therefore significantly under strength. See Ibid., pp. 83–84.

42 A. L. Dixon, 'The English police system', *Annals of the American Academy of Political and Social Sciences*, 1929, 146: 177–192, p. 179.
43 Laybourn and Taylor, *Policing in England and Wales*.
44 Ambage and Clark, 'Unbuilt Bloomsbury', p. 297.
45 HMSO, *Home Office: Report of the Advisory Committee on the Scientific Investigation of Crime*, London: His Majesty's Stationery Office, 1936 (hereafter HMSO, *Scientific Advisory Committee*); HMSO, *Detective Committee*.
46 Ambage and Clark, 'Unbuilt Bloomsbury', p. 302.
47 Ibid., p. 304.
48 Ibid., p. 305.
49 HMSO, *Scientific Advisory Committee*, Vol. 1, p. 5.
50 Ibid., p. 6.
51 *Medico-Legal and Criminological Review* became the *Medico-Legal Journal* in 1947.
52 HMSO, *Detective Committee*, Vol. 1, pp. 9–10.
53 Ibid. Also see Maxwell, 'The English police system', p. 87.
54 R. M. Howe, *Criminal Investigation A Practical Textbook for Magistrates, Police Officers and Lawyers*. London: Sweet and Maxwell, 1949, 4th edition. p. v.
55 N. Ambage, *The Origins and Development of the Home Office Forensic Science Service, 1931–1967*, unpublished PhD thesis, Lancaster University, 1987.
56 A. L. Dixon, *Atomic Energy for the Layman*, London: Chantry Publications,1950.
57 My own wide search for the term 'forensic science' in periodicals reveals very few instances where the term was used in the UK before the 1930s.
58 For Detective Committee, see Laybourn and Taylor, *Policing in England and Wales*, pp. 82–83. For Tryhorn's scientific aids papers, see Tryhorn, 'The packing of exhibits'; F. G. Tryhorn, 'The assessment of circumstantial evidence', *The Police Journal*, 1935, 8(4): 401–411; F. G. Tryhorn, 'Scientific aids in criminal investigation Part I', *The Police Journal*, 1936, 9 (1): 33–41; F. G. Tryhorn, 'Scientific aids in criminal investigation. Part II Searching at the scene of crime', *The Police Journal*, 1936, 9 (2): 152–160; F. G. Tryhorn, 'Scientific aids in criminal investigation. Part III Searching and packing', *The Police Journal*, 1936, 9 (3): 303–317; F. G. Tryhorn, 'Scientific aids in criminal investigation. Part IV Dust', *The Police Journal*, 9 (4): 403–412. Tryhorn also contributed to the Forensic Science Circulars. See F. G. Tryhorn, 'The fracture of glass', *Forensic Science Circular*; 1936, No. 2, London: HMSO, 1–5.
59 Laybourn and Taylor, *Policing in England and Wales*, p. 82.
60 J. Ward, *Origins and Development of Forensic Medicine and Forensic Science in England, 1823–1946*, unpublished PhD thesis, The Open University, 1993; Ambage, *The Origins*.
61 Volume 1 covered the police system, state of crime and review of detective work; Volume 2, the selection and training of detectives; Volume 3, crime records; Volume 4, communications; Volume 5, the application of science to the investigation of crime.
62 Ambage, *The Origins*.
63 HMSO, *Detective Committee*, Vol. 1, p. 16.
64 Ibid., p. 19.
65 Ibid. p. 22.
66 Ibid. p. 44.
67 Ibid., Vol. 2, pp. 20–22.
68 Ibid., Vol. 2, p. 22.

69 HMSO, *Home Office: Scientific Aids to Criminal Investigation: Instructional Pamphlet for the Use of Police Officers*, London: His Majesty's Stationery Office, 1936, (subsequent editions: 1949/1950, reprinted 1964).
70 HMSO, *Detective Committee*, Vol. 2, pp. 26–46.
71 Ibid., Vol. 5, p. 4.
72 HMSO, *Home Office, Scientific Aids to Criminal Investigation, Forensic Science Circulars*, No. 1, London: HMSO, 1936; HMSO, *Home Office: Scientific Aids to Criminal Investigation: Instructional Pamphlet.*
73 HMSO, *Detective Committee*, Vol. 5, pp. 16–19.
74 See A. Dunlap, 'Science versus practical common sense in crime detection', *American Journal of Police Science*, 1931, 2 (4): 322–327, for a critique of college-trained detectives.
75 HMSO *Detective Committee*, Vol. 5, p. 7.
76 Ibid., Vol. 5, p. 8.
77 Ibid., Vol. 5., p. 10.
78 Ibid., Vol. 5, pp. 8–9.
79 Ibid., Vol. 5, pp. 10–11.
80 Ibid.
81 Ibid., Vol. 5, pp. 12–13.
82 Ibid., Vol. 5, p. 14.
83 Ibid., Vol. 5, p. 10.
84 C.R.M. Cuthbert, *Science and the Detection of Crime*, London: Hutchinson, 1958, p. 9.
85 H. J. Walls, 'The Forensic Science Service in Great Britain: A short history', *Journal of the Forensic Science Society*, 1976, 16: 273–277, p. 274.
86 35s = £1.75 which would be equivalent to around £100 in 2015 which is not expensive for a microscope.
87 Cuthbert, *Science and the Detection of Crime*, p. 11.
88 Walls, 'The Forensic Science Service', p. 274.
89 Ibid., p. 275.
90 Ambage, *The Origins*, p. 81.
91 Ibid., p. 72.
92 Ambage and Clark, 'Unbuilt Bloomsbury', p. 299.
93 Walls, 'The Forensic Science Service', p. 275.
94 Ambage and Clark, 'Unbuilt Bloomsbury', p. 299. In his autobiography, Glaister stated that he was approached with the suggestion that he 'should take over the medical directorship of the Metropolitan Police College at Hendon' which must mean the directorship of the laboratory which was not, of course, a medical role. See J. Glaister Jnr., *Final Diagnosis*, London: Hutchinson, 1964, p. 119. Glaister gave his reasons for not pursuing this as the suggestion, made by Lord Tenchard, that he was considering militarizing the Metropolitan force which would have meant restrictions and the fact that he had better facilities for research at Glasgow.
95 Ambage, *The Origins*, p. 71.
96 Walls, 'The Forensic Science Service', p. 275.
97 Ibid., p. 276.
98 From British Pathé, *Burglars Beware* (1935), available at http://www.britishpathe.com/video/burglers-beware-aka-burglars-beware/query/Hendon+laboratory, accessed 6 May 2015. Image reproduced by kind permission of British Pathé.

99 H. J. Walls, *Expert Witness: My Thirty Years in Forensic Science*, London: John Long, 1972, p. 15.
100 Ibid., p.14.
101 M. Pereira, 'Dr Henry James Walls', *Journal of the Forensic Science Society*, 1988, 28 (5–6): 357–358.
102 Ambage, *The Origins* p. 85.
103 Ambage and Clark, 'Unbuilt Bloomsbury'.
104 Ambage, *The Origins*, p. 85; also see Walls, 'The Forensic Science Service', p. 275.
105 Walls, 'The Forensic Science Service', p. 277.
106 Ibid., p. 276.
107 Walls, *Expert Witness*, p. 16.
108 Ambage, *The Origins*, p. 95.
109 Figure 4.2 and 4.3 are from British Pathé, *Science Fights Crime* (1946). Reproduced by kind permission of British Pathé.
110 British Pathé, *Science Fights Crime* (1946), available at http://www.britishpathe.com/video/science-fights-crime/query/for, accessed 5 May 2015.
111 Ambage, *The Origins*, p. 92.
112 Ambage and Clark, 'Unbuilt Bloomsbury', p. 299; Ambage, *The Origins*, p. 93.
113 Ambage, *The Origins*, p.104.
114 Ibid., p. 103.
115 HMSO *Detective Committee*, Vol. 5, p. 66.
116 Else and Garrow, *The Detection of Crime*.
117 Ibid., p. 26.
118 Ambage, *The Origins*, p. 105.
119 Ibid., p. 108.
120 Ibid., p. 114.
121 Firth, *A Scientist*.
122 Ibid., p. 18.
123 Ibid., p. 19.
124 Ibid.
125 'Scientific aids to police work: criminals traced by analysis', *The Manchester Guardian*, 28 July 1939, 3.
126 Firth, *A Scientist*, p. 22.
127 Ibid.
128 'Scientific aids', *The Manchester Guardian*, 28 July 1939, 3.
129 Ibid. quoting Dixon.
130 Ibid.
131 'Scientific institute to help the C.I.D.', *The Daily Mail*, 11 April 1935, 4.
132 The *Daily Mail* article (11 April 1935) was eighty lines long and a single column width (without the initial headlines). About fifty-six lines described the call for a medico-legal institute and attendant functions: twenty-four lines described the functions of the new police laboratory. Lord Atkin was a senior judge and a former president of the Medico-Legal Society (see Ambage, *The Origin*, p. 74). Atkin was a member of the Scientific Advisory Committee.
133 'Scientific institute', *The Daily Mail*, 11 April 1935, 4.
134 Walls, *Expert Witness*.
135 Lord Atkin mentioned Sherlock Holmes *and* John Thorndyke in his opening speech. 'It gave him great satisfaction to think that Sherlock Holmes and Doctor Thorndyke, of whom they might have heard, were to be succeeded in the eyes of the world by

Dr. Davidson, the Head of the Laboratory', Anon., 'The opening ceremony of the Metropolitan Police Laboratory', *Metropolitan Police College Journal*, 1935, 1:19–21, p. 20.

136 British Pathé, *Burglars Beware* (1935).

137 Ambage, *The Origins*, p. 78.

138 Before the advent of television news, British Pathé newsreels were important rivals to newspapers for news. See http://www.britishPathé. com/pages/history (accessed 6 May 2015).

139 Ambage, *The Origins*, p. 120.

140 'Scientific aids to police work: Criminals traced by analysis', *The Manchester Guardian*, 28 July 1939, 3.

141 Ambage, *The Origins*, p. 121.

142 Ibid.

143 Ambage, *The Origins*, p. 306.

144 M. Lynch, 'Science, truth, and forensic cultures: The exceptional legal status of DNA evidence', *Studies in History and Philosophy of Biological and Biomedical Sciences*, 2013, 44 (1): 60–70.

145 Ibid., p. 60.

146 Ibid., p. 68.

147 HMSO, *Forensic Science Circulars*; HMSO, *Instructional Pamphlet* (see note 69); HMSO, *Detective Committee*; Else, and Garrow, *The Detection of Crime;* for *Metropolitan Police College Journal* see, e.g., C. T. Symons, 'Scientific aids in prospect', *Metropolitan Police College Journal*, 1935, 1 (2): 30–34. The journal ceased publication in 1939, presumably because of the outbreak of war. For Tryhorn's *Police Journal*, 'scientific aids' articles, see note 58.

148 L. C. Nickolls, *The Scientific Investigation of Crime*, London, Butterworth, 1956; Cuthbert, *Science and the Detection of Crime*.

149 Lucas, *Forensic Chemistry*; C. A. Mitchell, *Science and the Criminal*, London: Isaac Pitman & Sons, 1911; C. A. Mitchell. *The Scientific Detective and the Expert Witness*, Cambridge: W. Heffer & Sons Ltd, 1931; C. A. Mitchell, *A Scientist in the Criminal Courts*, London: Chapman & Hall Ltd, 1945; J. Grant, *Science for the Prosecution*, London: Chapman and Hall, 1941.

150 HMSO, *Instructional Pamphlet*, p. 3.

151 Ibid., pp. 3–4.

152 Police Journal, 'Editorial', *The Police Journal*, 1928, 1 (1): 1–2.

153 Ibid., p. 2.

154 Ibid.

155 J. Glaister Jnr., 'Some results of recent medico-legal research in the examination of blood-stains and hairs', *The Police Journal*, 1928, 1 (1): 62–77; E. Locard, 'Dust and its analysis: An aid to criminal investigation', *The Police Journal*, 1928, 1(2): 177–192; C. A. Mitchell, 'Science as applied to circumstantial evidence', *The Police Journal*, 1928, 1 (2): 256–267; H.T.F. Rhodes, 'The nature of chemical evidence and its place in the detection of crime', *The Police Journal*, 1928, 1(4): 657–668; S. Smith, 'The identification of firearms and projectiles: As illustrated by the case of the murder of Sir Lee Stack Pasha', *The Police Journal*, 1928 1(3): 411–422.

156 W. Willcox, 'Toxicology and crime', *The Police Journal*, 1928, 1 (1), 98–104.

157 A. L. Dixon, 'The new series of the Police Journal', *The Police Journal*, 1933, 6 (1): 1–3.

158 Ibid. p. 2.

159 See note 58 for a list of Tryhorn's 'scientific aids' articles.

160 Tryhorn, 'The packing of exhibits'.
161 E.g., see N. Kendal, *Criminal Investigation A Practical Textbook for Magistrates, Police Officers and Lawyers*, London: Sweet and Maxwell, 1934, 3rd edition, p. 297.
162 A. Lucas, *Forensic Chemistry and Scientific Criminal Investigation*, London: Edward Arnold, 1931, 2nd edn.
163 Lucas, *Forensic Chemistry and Scientific Criminal Investigation*, pp. 14–15.
164 J. McCafferty, *Mac, I've Got a Murder*, London: Arthur Barker Ltd., 1975, p. 26.
165 Ibid., p. 82.
166 Ibid., p. 83.
167 Ibid.

5 Forensic science careers and self-images

Introduction

The new forensic scientists, who worked as independent forensic experts or who took up the new roles in forensic science laboratories in the 1930s and 1940s, were not generally well-known public figures, although, as today, involvement with high-profile murder cases tended to bestow a level of public recognition through media attention. Additionally, some found themselves in the public eye quite late in their careers when they became associated with a prominent case.[1] This was the first generation of forensic scientists, although they would not necessarily at this stage have designated themselves thus; rather, they identified themselves in terms of their parent disciplines of chemistry, biology and physics.

Walls described himself as a physical chemist, and in terms of his initial employment in the Metropolitan Laboratory, a 'physicist'.[2] Nevertheless, by badging its network of laboratories as forensic science laboratories and through using the words 'forensic science' in its various publications, the Home Office promoted the use of the terms 'forensic science' and 'forensic scientist'. These scientists were civil servants employed in the Home Office laboratories, or were independent scientific consultants who undertook forensic work as part of their professional portfolios, the latter being a much longer established role. Most had, by one means or another, developed expertise in sciences relevant to crime detection through university work, industry, consultancy or public analysis. This period witnessed a move towards the professionalization of forensic science with its delineation from forensic medicine.

Having already considered the setup of forensic science laboratories and the rhetoric of the 'scientific aids' movement, this chapter considers the question of the professionalization of forensic science through alternative routes into forensic work that some of the new breed of forensic scientists travelled in their careers, starting with the roles of public analyst and Home Office Analyst. The forensic careers of public analysts and analytical chemists involved in forensic work represented an earlier type of forensic position in comparison to the Home Office laboratory roles and a kind of parallel route into forensic chemical work through independent consultancy; in the early days, medical and chemical expertise were strongly linked. The following section considers the separation of forensic science and forensic medicine and what this implied for professionalization. Finally, biographical material is analysed to shed light on the lives of the new forensic scientists. Autobiography, in particular, reveals important aspects of how scientists wished to portray their self-images.

Home Office Analysts, public analysts and the Government Laboratory

The development of the forensic science profession was complex and piecemeal. From the end of the nineteenth century and into the twentieth century, there were several medical and scientific roles requiring knowledge of chemical analysis for forensic purposes. Undoubtedly, criminal poisoning cases provided a considerable stimulus to the development of forensic medicine and associated chemical analyses. From the mid-eighteenth century until the beginning of the twentieth, while pathology was still developing as a specialism, 'other scientists, often chemists were integral to the resolution of the particular crime of poisoning'.[3] By the middle of the nineteenth century there was a growing group of expert toxicologists, a situation which was fuelled by the Society of Apothecaries' requirement that candidates for their diploma attend lectures in medical jurisprudence.[4] Typically, in suspected poisoning cases, the first doctor available undertook the post-mortem, sending samples and organs to a suitable chemist for analysis; the chemist only stated the quantity and nature of the poison if it was found to be present; cause of death was the province of the medical doctor.[5] By the mid-nineteenth century the involvement of professional chemists and toxicologists was consolidated, only to be superseded by the contribution of public analysts. Hospital pathologists and police surgeons were also undertaking post-mortems later in the century.[6] Chemistry lecturers in medical schools could be called upon for toxicological analyses, and some surgeons had skills in chemistry.

From the 1880s, the role of Director of Public Prosecutions in England provided additional stimulus for the Home Office's interest in scientific support for criminal prosecutions, which was to culminate in the opening of forensic science laboratories from the 1930s.[7] The scientific employees of these laboratories were, in effect, salaried civil servant forensic scientists. In that respect they were quite privileged, as most roles involving such scientific analyses were sustained by individuals working on a fee basis until the mid-1930s. Although forensic medicine was a much longer established discipline, the Home Office was late in recognizing the need to appoint state forensic medical specialists. The push to create such roles came from the continuing problem of suspected poisonings, hence the need to have qualified analysts who could advise the Home Office. The position of Home Office Analyst, which included work in forensic medicine, analytical chemistry and toxicology, was created in 1882 with the appointment of Thomas Stevenson as the first Scientific Analyst to the Home Office.[8] The Analyst was often called as the main expert witness in problematic murder cases and could be asked to examine clothing and crime scenes, the latter areas falling under the umbrella of what would eventually become forensic science.[9] Stevenson, a remarkably energetic individual, was already a public analyst and a medical officer of health when he took up the Home Office Scientific Analyst role, 'a position that was to involve him in nearly all the most notorious poisoning cases of the next forty years'.[10]

The brief for the Home Office Analyst quickly became wider than that of toxicology. As a laboratory-based medical scientist, this role was at the elite end of forensic medicine. Indeed, Ward contends that when the Metropolitan Police Laboratory was being set up, 'it was this elite laboratory-based practice of forensic medicine which served as a model for the new "forensic science"'.[11] This may have been part of the reason why medico-legal experts were consulted on the thorny issue of the Directorship of the Metropolitan Police Laboratory before its opening in 1935. The Home Office Analyst role helped redefine forensic medicine as medico-legal analyses for police purposes.[12]

Although there were qualified chemists who could have held Home Office Analyst posts, these positions usually went to those who were medically qualified. The tendency to appoint those with medical degrees was reinforced by the involvement of the medical royal colleges; hence, analytical chemists without medical qualifications were likely to be overlooked for such appointments.[13] Even so, knowledge of hospital chemistry was clearly an important part of the role. Home Office Analysts were trained through a system that was effectively an apprenticeship. The master passed on his knowledge to the next generation, who passed it on to the next, and so on, so there was an unbroken chain of apprenticeship stretching for a century which transferred the skills needed to appear for the prosecution and withstand cross-examination. 'The creation of an elite branch of forensic medicine as Home Office analysis, then, depended on the hospital chemistry laboratory.'[14] William Willcox became Home Office Senior Analyst in 1908. Willcox trained up Gerald Roche Lynch who was Senior Analyst from 1927 to 1954.[15] Roche Lynch was, of course, trained in chemistry and medicine and was active over the period when the forensic science laboratories were set up; indeed, Arthur Dixon had his eye on Roche Lynch as a contender for the Director of the Metropolitan Police Laboratory at one point.[16]

The Home Office appointed special pathologists, notably, Bernard Spilsbury. He worked in this role with Willcox in St Mary's and enjoyed considerable status. At the bottom of the pecking order came the police surgeons.[17] However, for the elite Home Office Analysts and pathologists, their professional remuneration belied their status, as they had to rely on fees rather than a salary (and there was no pension in retirement). Spilsbury's Home Office appointment was honorary until 1922. He only received individual fees for each case plus a modest amount for the upkeep of his lab.[18] Willcox as Home Office Analyst combined the role with lecturing and hospital duties.

Food adulteration was endemic in the mid-Victorian period.[19] From the 1860s to the 1890s a series of acts was passed in an attempt to reduce the problem. The Adulteration of Food and Drink Act (1860) created the public analyst role and allowed for the appointment of public analysts by local authorities, although there was little interest until their appointments became compulsory.[20] Only after the Sale of Food and Drugs Act (1875) did local authorities begin to take seriously the appointment of public analysts. The Society of Public Analysts and Other Analytical Chemists was set up in 1874 to represent the new professional role.[21] The society and its journal, *The Analyst*, were important professional homes for early forensic scientists, especially those who were independent analytical chemists such as Charles Ainsworth Mitchell. From the start, these scientists needed to hone their courtroom skills, skills which were quite different from the analytical skills required in the laboratory: 'articulacy skills of the witness are very different from those of a laboratory scientist, so this required a hybrid expertise that was performative'.[22]

Although public analysts were appointed to investigate food and water adulteration, many were interested in toxicology and began to appear in toxicology trials; some had been doing so before they were appointed as public analysts.[23] A number of public analysts were qualified in medicine.[24] 'Those who filled these new public positions quickly became sources of expertise to other branches of state apparatus, including coroners and the police.'[25]

The Excise Laboratory, founded under the 1842 Tobacco Act to prosecute those committing fraud, was the first state laboratory to undertake forensic analyses, but the emphasis was on extracting revenue rather than the interests and well-being of the public.[26] The laboratory employed analytical chemists who had undertaken chemical training either at University College, London, or the Royal College of Chemistry.[27] With various mergers, re-organizations and changes of name, the laboratory became the Board of Inland Revenue Chemical Laboratory (1849–1894), the Government Laboratory (1894–1911), Department of the Government Chemist in 1911 and was then absorbed into the Department of Scientific and Industrial Research (DSIR) in 1959, becoming the Laboratory of the Government Chemist.[28] Almost from the start the laboratory was involved in testing a wider range of substances than tobacco, e.g., tea, pepper, soap and beer, greatly helped by its group of trained analytical chemists.[29]

Unfortunately, the Inland Revenue Laboratory was put in contention with the public analysts because the 1875 act, in addition to giving public analysts more powers, required central referees to be appointed to carry out tests in disputed analysis cases.[30] The Inland Revenue Laboratory was proposed as the refereeing body. Although local authorities had to employ public analysts, the effect was that their analyses, whether for prosecution or defence, could be challenged in a court of law, and the Inland Revenue Laboratory could be asked to do its own analyses to contest the public analysts' results.[31] This hardly made for a harmonious working relationship amongst professional chemists and, at least in the 1880s, was a considerable source of grievance amongst public analysts, not least because most of the Inland Revenue Laboratory staff were not trained chemists, but were instead 'lay inspectors' with only enough chemical knowledge to undertake simple chemical manipulations, such as determining the strength of alcohol and similar operations. Although some Inland Revenue staff undertook a course in chemistry at the Royal College of Science, they did not take part in the activities of the Society for Public Analysts, and the public analysts viewed the laboratory as lacking appropriate scientific credentials.[32]

Steere-Williams argues that there were deep-seated epistemological and professional differences visible in conflicts between the public analysts and the government chemists in terms of analytical epistemology and 'the performative ways in which the results of analytical chemistry were communicated to larger British public, scientific, and social bodies, including legal ones . . . competing visions of analytical chemistry'.[33] The Society of Public Analysts was 'a means of protecting the interests of local public analysts against the newly appointed government laboratory, an antagonism that lay at the heart of its institutional ethos'.[34] It was a way of promoting the status of analytical chemists, to distinguish them from amateurs and from theoretical chemists who would be found in universities.

The refereeing powers of the Inland Revenue Laboratory ensured that relations between the laboratory and the public analysts remained poor for the rest of the century; these came to a head with an attack by Otto Hehner, Public Analyst for Derbyshire, the main spokesman for the public analysts, on the laboratory's methods and standards, centring on the analysis of milk.[35] The dispute was emblematic of competing views of the professional arena for analytical chemistry which were channelled into a fight both for public recognition and professional space.[36]

The laboratory continued to expand its work after World War I, and at the time when the forensic science laboratories were being set up, was an established centre for the analysis of a wide range of foodstuffs, beer, brewing materials, tea, tobacco, alcohol, water and household supplies for a variety of government departments, including the Admiralty, Ministry of Agriculture and Fisheries, the Crown Agent for the Colonies, Customs and Excise, Board of Trade and Office of Works.[37]

Colonial laboratories

Colonial state chemical laboratories were run on similar lines to the UK Government Laboratory. The Government Laboratory in Ceylon was directed by C. T. Symons, who returned to the UK to become the Home Office's Forensic Science Adviser, a post he held for a short time until his death in 1937.[38] Alfred Lucas trained at the Inland Revenue Laboratory, undertaking a formal course at the Royal College of Science, before moving to Egypt where he became Director of the Government Laboratory in Cairo until his retirement from that role in 1923.[39] The Cairo laboratory undertook a wide range of analyses of substances for the Egyptian government. The majority of its work involved chemical and physical inspection of materials supplied to the government 'to determine whether they are what seller describes, whether suitable for purpose, and whether worth the price'.[40] The major part of its police activities involved the examination of alcoholic liquor, and the laboratory routinely found that over 30 per cent of liquids were adulterated or wrongly described.[41] The laboratory had begun to undertake research by 1920 and police or 'chemico-legal' work as Lucas termed it, had expanded.

> The volume of work has increased very considerably during the last few years, but as yet only the fringe of what is required has been touched, although the value of chemistry as an aid in criminal investigation is becoming more and more recognized.[42]

The laboratory examined materials submitted to it by the Parquet (Public Prosecutor's Department of the Egyptian Ministry of Justice). Cases connected with poisoning and blood stains went elsewhere.[43]

> The work consisted in the examination of questioned documents for age, alterations, erasures etc., and occasionally the comparison of handwriting, the examination of counterfeit coins, and analysis of dust and dirt found in connection with a crime, the examination of clothes and textile fabrics of stains and marks other than blood and seminal stains, the recognition of fibres, the examination of firearms and bombs, the analysis of bullets and other projectiles for firearms, the inspection of premises in special cases of accident, fire, etc., etc.[44]

The Cairo laboratory dealt with cases of national importance. One example involved a claim against the Egyptian government for land valued at around 15 million Egyptian pounds. In connection with this, 168 documents were examined of which 163 were found to be forged. In a separate counterfeit coining case thirty-seven convictions were secured.[45] These were turbulent times in Egypt and this meant that the laboratory was often called into cases involving bullets and bombs:

> [A]nother case was in connection with the attempt made on the life of H.H. the late Sultan Hussein by a bomb thrown at his carriage; another was the examination of parts of the bombs in the attempted assassination of two successive Prime Ministers, namely H.E. Mohamed Said Pasha, and H.E. Wahnba Pasha, both of which attempts were fortunately unsuccessful.[46]

Lucas was reticent about political matters in Egypt, although his position as a prominent British official in a senior state role might have been problematic in the longer term, had he not retired in 1923. He did, however, appear to command a considerable degree of respect in Egypt as when he worked alongside Howard Carter on the Tutankhamun excavation, he was appointed not only as a chemist but as the Egyptian government's official representative on the excavation.[47] It was left to his colleague, Sydney Smith, in his role as Medico-Legal Expert in Cairo to describe the dangers, particularly to the British in the years before Egyptian independence, noting the occasion sometime between 1919 and 1922 when anti-British feeling was at its height and he took refuge on the roof of the Government Laboratory from the riots taking place in the streets below. In addition to its civilian roles of testing and advising on analyses of materials, the Government Laboratory in Cairo was responsible for the kinds of analyses which, in the UK, fell within the sphere of the public analyst or (some years later) to the forensic science laboratory.

The Analyst regularly published a section: 'Notes from the Reports of Public Analysts' where domestic and colonial reports were published. The following year the Cairo lab reported:

> Investigations were made in connection with 62 cases involving firearms, projectiles and bombs (including the examination of the fragments of bombs thrown at four different Ministers during the year), 17 seal impressions, 75 questioned documents, 9 forged banknotes, and 5 cases of counterfeit coins (305 articles).[48]

In the early part of the twentieth century, Egypt's Medico-Legal Advisers in the form of Sydney Smith and then John Glaister Jnr continued their careers in Scottish universities, having cut their forensic teeth in the well-equipped medico-legal laboratory in Cairo. Lucas's personal interest in the forensic aspects of chemical analyses is demonstrated through the four editions of his work, *Forensic Chemistry*. He continued as an independent forensic adviser after his retirement, continuing in this role alongside his archaeological work. Although Lucas was never a public analyst – almost all his professional career was located in Egypt – nevertheless his professional sensibilities belonged in public analysis, as so much of his Government Laboratory work in Cairo involved similar types of analyses; his analyst roots are clearly visible in his textbook. He frequently referred to himself as an analyst, by which he meant an analytical chemist, in his writing, and he published in the journal *The Analyst*, corresponding with other British forensic analytical chemists such as Charles Ainsworth Mitchell.[49]

Chemical dynasties

In the UK, the staff of the Department of the Government Chemist, as it was called in the 1930s, were frequently called upon to analyse a wide range of substances, so they often appeared in court in an expert witnessing role which gradually developed over the lifetime of the laboratory. For instance, in 1935 court cases involving laboratory staff as expert witnesses included 'prohibited drugs, alcohol in blood, tobacco and saccharin smuggling, illicit spirits, beer adulteration, the illegal admixture of petrol and kerosene, salmon poaching and the watering of milk and the abstraction of cream'.[50]

Most of the work of the Government Chemist was for government departments; however, public analysts were the front-line chemists employed by local authorities, analysing samples of foodstuffs and other materials to detect potential adulteration. Some made a career of public analysis, held several public analyst roles, became active in the society and trained up successive generations of public analysts. As well as having similar toxicological and analytical interests, the apprenticeship aspect of the public analyst role was similar to that of the medically trained Home Office Analysts, so there was also, in effect, a dynasty of analytical chemists. One such dynasty was spawned through the influence of Otto Hehner (1853–1924). It is helpful to give some detail of Hehner's career, not so much because he was directly influential in the lead-up to the inauguration of the forensic science laboratories (he was not); rather, he exerted considerable influence on the next generation of public analysts and independent analytical chemist expert witnesses. His influence spanned a generation and more of independent forensic scientists, including Charles Ainsworth Mitchell and Julius Grant.

Hehner was born in Germany and received his initial chemical education there, coming to the UK and taking up an appointment as an assistant at Andersonian College, Glasgow, in 1873 where he worked alongside the chemist William Ramsay, the discoverer of five noble gases.[51] He was invited to work with Hassall, the main authority on the chemistry of food, began to research animal fats and became a scientific authority on butter.[52] He set up his practice as an analytical and consulting chemist in London in 1877 and was appointed as public analyst for various authorities, positions he held from 1878 to 1922. He was an active member of the Society for Public Analysts, becoming President in 1891. He researched widely on preservatives, beer, vinegar, milk, butter, water and beeswax, becoming a Fellow of the Institute of Chemistry, serving as Vice-President three times and serving as an examiner for them. He vocally attacked the use of colourings and preservatives and was part of a movement of food chemists who questioned the so-called health benefits of food that had been subject to what were described as scientific treatments. Hehner was one of the leading public health chemists of the day and served on several Parliamentary committees. During the First World War he was the official referee between the War Office and soap manufacturers for glycerin contracts. He was an enthusiastic expert witness in the law courts. As Ainsworth Mitchell noted, he was

> . . . never happier than when he was working up the chemical side of a case and devising experiments which could be shown in Court to convince judge and jury of the truth of his contention . . . an excellent witness, clear and concise in his evidence, and always ready with a humorous answer for a cross-examining counsel.[53]

So, not only was Hehner directly influential in the careers of a number of public analysts and analytical chemists, his example also acted as a role model for how such scientists could effectively present evidence and perform in court.

One of Hehner's most illustrious protégés was Charles Ainsworth Mitchell. Perhaps, the most succinct description of Mitchell's career was as a 'private public analyst'. Although Mitchell does not appear to have held any official public analyst appointments, much of his career was spent on the analysis of food products and many of his close professional associates were public analysts and analytical chemists. After graduating in chemistry from Oxford, he worked as an assistant to Otto Hehner.[54] Mitchell worked at Beaufoy's vinegar brewery from 1897 to 1932, possibly the first qualified chemist to work in the industry, before becoming a consultant when it became British Vinegars Ltd. As well as writing many scientific articles, he was a prolific author of books on food: oils, fats, vinegar, mineral waters and fermentation.[55] He gave evidence on a number of spy trials in World War I in connection with postal censorship.[56] Notably, in 1915 he was called into an espionage trial involving the question of whether lemon juice had been used as invisible ink. This involved demonstrating that a pen had been corroded by lemon juice and that acid in the lemon juice contained iron. Apparently the suspect lemon is still filed in the National Archives, although its ability to furnish invisible ink has long since evaporated.[57]

Alongside state roles as public analysts or analytical work for other state bodies, many analysts undertook consultancy work for industry. Horrocks shows that chemistry consultancy in the food industry was widespread in the later nineteenth and into the twentieth century.[58] Firms would use a consultant as a way of claiming the superior quality of a product, and this was something that could be put to use in advertising.[59] It was Mitchell who introduced public analyst and food chemist Henry Cox to Hehner and this resulted in the setup of the practice of Hehner and Cox in London in 1923.[60] After Cox's death in 1951, Hehner and Cox was led by Julius Grant who himself had trained under Hehner. Grant ran Hehner and Cox as a chemical consultancy well into the 1980s, whilst also holding an appointment in the paper industry. Arguably there were echoes of Hehner's influence visible in analytical and forensic chemistry consultancy for a hundred years.

Hence, analytical chemists of the late nineteenth and early twentieth centuries often had wide-ranging and long-standing professional networks to draw upon, encompassing those they had trained with or under and those they worked alongside in publishing and consultancy. With the example of William Ramsay, who worked with Hehner and other consulting chemists, Watson argues that a system of internal referral must have arisen among a small group of fairly elite scientists.[61] The extensive networks of analytical chemists from at least the late nineteenth century and well into the twentieth, who were used to presenting evidence in court cases, demonstrate that public analysis and analytical chemistry were fertile training grounds for new independent consulting forensic scientists.

Two of the earliest English-language textbooks on forensic chemistry, William Jago's (1909) *A Manual of Forensic Chemistry* and Alfred Lucas's (1921) *Forensic Chemistry*, reflect the public analyst route into forensic work.[62] Jago's book was based on a course of lectures on forensic chemistry delivered at University College, London. This text is remarkable, as it spans the traditional concerns of the public analyst with substantial chapters on adulteration of food and drugs, new manufacturing processes and use of preservatives and colouring substances, before discussing criminal matters in terms of death by poisoning or drugs halfway through the book, and then returning to ways of dealing with chemical evidence in the court in considerable detail in terms of civil actions and food and drugs legislation. The book focuses on evidence in general and, importantly, describes how the scientist should proceed in the courtroom with regard to representing evidence; the book is much less about the actual chemistry itself. The importance of Jago's work, now largely forgotten, lies in its understanding of the link between the work of the public analysts and the understanding that scientists needed to know how to handle their appearances in court. As Chapter 1 argues, the latter was an aspect of the forensic scientist's work which was a real concern to incumbents of professional forensic roles, as it was not a skill which came naturally to all scientists.

William Jago (1853–1938) trained under Edward Frankland at the Royal College of Science and Royal School of Mines before becoming a school science teacher. He developed an interest in brewing and the science of bread making, publishing books on the science and technology of wheat and bread making.[63]

After many court appearances, he developed a reputation as a scientific expert witness; this spurred him on to become qualified in law and was called to the bar in 1904.[64] His book on forensic chemistry amplified his public analysis interest and his knowledge of the problems which the scientist would face in courts of law. Jago's book was not as widely referenced, as were the later successive editions of Lucas's book on forensic chemistry. As such, Lucas's work focused more on criminal aspects, yet it also discussed the concerns of the analyst in terms of alcohol, building materials and food and drugs. Although Jago's career was at its peak rather before the first flowering of forensic science in England and Wales, nevertheless his work is an important precursor of the discipline and reinforces the link between chemical analysis, the public analyst role and forensic work, and highlights the continued need for forensic scientists to develop their courtroom skills alongside their scientific skills.

Forensic science or forensic medicine?

The 1930s were an important period for the professionalization of forensic science. One of the markers of professionalization is demarcation from other professions. In this case, a key boundary was increasingly being drawn between forensic science and forensic medicine. As the preceding sections have argued, medical and chemical issues were thoroughly intertwined through the work of the Home Office Analysts and public analysts; the requirements of forensic toxicology made this combination of skills desirable. In the twentieth century, forensic pathologists often had broad interests across forensic matters. For instance, as Chapter 1 describes, it is notable that a clutch of eminent medico-legal specialists, including Spilsbury (with the help of Churchill the gunsmith), Smith, Littlejohn and Glaister, were all involved either in undertaking ballistics experiments or advising on such experiments.[65] There were professional demarcations; post-mortems and other clearly medical matters could only be undertaken by medically qualified individuals. Hence, when analytical chemists such as Lucas and Mitchell were advocating the professional autonomy of forensic chemistry, they were arguing for the separation of subjects which had hitherto been inextricably linked in the forensic arena.

In his autobiography, Sydney Smith reflected on the specialization of forensic medicine and how public health, once included in the discipline, had long since been separated. Forensic pathologists, forensic serologists and forensic toxicologists were now all specialists in their own right, he argued.[66] He recognized that not all of forensic science was medical and advocated cooperation between botanists, zoologists and so on. Smith recognized that the boundaries between forensic disciplines had been fluid and that there were some 'anomalies.' Notable amongst these was ballistics, as there was nothing medical about the study of bullets and cartridge cases. As there was no separate discipline of forensic ballistics, Smith claimed, during his career, such analyses had come to the medico-legal expert: 'because he was accustomed to dealing with bullets in relation to wounds, and because there was no one else to undertake it'.[67] Later in his autobiography, Smith alluded to Lucas's interest in ballistics at the Government Laboratory in Cairo, noting that Lucas believed it should be a branch of forensic chemistry.[68] Lucas was, of course, a colleague of Smith's in Cairo, and they often exchanged information on shooting cases.[69] Smith was able to claim the area as part of his domain as: 'Bullets were our concern once they entered the body and it seemed reasonable to find out from them all we could'.[70] In fact, Smith gained a considerable degree of professional acclaim for his work on forensic ballistics in his role of Medico-Legal Expert in Cairo, through his identification of the murder weapons in the shooting of Sir Lee Stack, head of the Egyptian army (Sirdar) and Governor General of Sudan, in 1924.[71] This was an immensely important political murder and also one of the first times that the gun used in a murder had been definitively identified by markings on the bullets.

It is interesting that the professional line of demarcation between Lucas, the forensic chemist, and Smith, the medico-legal expert, should focus on forensic ballistics which, to a modern eye, falls into neither area. But claims to professional authority were often more than claims that a particular subject belonged to a particular scientific discipline – they were claims to the moral authority of the scientist. Chapter 4 described Lucas's arguments that chemists were well qualified to deal with the minutiae of forensic packing. But he made a more general plea for the value that the chemist could offer to the criminal justice process based on character, a quality beyond chemical knowledge alone.

> The chemist is specially trained in observation and deduction; he learns to enquire closely into the nature and meaning of things; as a result of experience he becomes very sceptical of outward appearances; if he practises as an analyst he is brought into intimate contact with a great variety of different materials, of all of which he must know the composition and properties, and he deals habitually with traces of substances and weighs and measures them, all of which qualifications peculiarly fit him to act as a scientific expert in criminal investigation. Observation, deduction, an enquiring turn of mind, a scepticism of appearances, a knowledge of the composition, properties and uses of materials and the ability to see and examine traces are all essential for this work and it may be stated confidently that if criminal investigation is to keep abreast of the times it must be made more scientific and chemical analysis and microscopy must be used more frequently.[72]

By way of contrast, Smith ascribed similar qualities to the specialist in forensic medicine:

> [T]he part played by the medico-legal expert is not so circumscribed as that, for example, of an analytical chemist helping the police. He sees the case as a whole, not exclusively in its medical aspects. He observes, infers, and even speculates. To him, because of his special knowledge, a non-medical clue may have a significance that even an astute police officer has not grasped. His peculiar experience and talents may enable him alone to deduce the correct interpretation of the facts.[73]

Nevertheless some professional tensions were evident, and these often centred on the ability of forensic chemists to delineate their professional role. Writing in 1938 Mitchell described uncertainties over the scope of forensic chemistry because of the tendency for forensic toxicology to be regarded as part of forensic medicine. Some, he argued, even regarded forensic medicine as covering the whole ambit of sanitation, food analysis, forgery and counterfeiting and 'all criminal offences requiring chemical evidence'.[74] It was important to demarcate between chemical matters that required medical interpretation and those completely outside the scope of medicine. Mitchell was critical of the initial setup of the public analyst system in 1872 where local authorities 'frequently took the easy course of thrusting their medical officers into the posts'.[75] This resulted in all sorts of workarounds to undertake the chemical analyses required. The issue only began to be resolved after 1900 when public analysts were required to have competence in microscopy and qualifications in analytical chemistry. Mitchell was similarly critical of the Scientific Advisory Committee's (1936) report which advocated the setup of a medico-legal institute.

> It will be seen that in this report the tradition of the subordination of the chemist to the medical man is unconsciously perpetuated. The scheme . . . is far too restricted . . . What is required is an Institute for Forensic Medicine and Chemistry, in which the two professions can collaborate on a basis of equality.[76]

However, it would be wrong to exaggerate the tensions that might have existed as forensic scientists attempted to carve out their disciplines. Grumblings about the state of one's professional circumstances often related more to a general lack of resources rather than to a charge that some other professional group was unfairly encroaching on matters outwith its domain and commanding an unfair share of resources. Forensic scientists and pathologists worked alongside each other on many cases and also occupied a common professional space in professional societies, which were meeting grounds for specialists in forensic medicine and forensic science.

However, there were times when the requirement for a forensic scientist on the case was vital, even in murders where medical expertise took centre stage. Firth gave an example as to why an appropriate specialist forensic scientist would often be needed from an example early in his career at the North Western Laboratory.[77] The case involved a woman's body found in a field. In those days, the usual course of action was for the coroner to call in the police surgeon. In this case the police surgeon assumed she had died of exposure – he was not keen to look for problems – and the coroner did not want to oppose this judgment. However, Firth suspected she had been strangled. Some tiny objects in the woman's nose were a potential clue; the police surgeon thought they were seeds from a tree, Firth knew these were bluebottle eggs. He removed them and measured the length of time until they hatched. The incubation time of twelve to fifteen hours could have been extremely important in helping to establish the limits of the time when the crime could have been committed: 'a good example of the sort of tiny clue which only the expert can make use of'.[78] Although Firth pointed out a piece of evidence which could have been vital, and it did not appear to have been a difficult experiment to conduct, nevertheless he was a chemist sailing into biological waters. This serves as a further illustration that, apart from matters unequivocally medical, forensic scientists and medico-legal specialists were guilty of many raids into scientific specialisms beyond their own in regard to the analysis of forensic evidence.

It was not the case that forensic pathology was well provided for and forensic science was a Cinderella subject in terms of resources. Far from it – all forensic areas appeared to be under-resourced before and after World War II. Firth regarded the medical situation in the early days of his forensic career as very problematic, as there were few people qualified to carry out a post-mortem.[79] Even skilled hospital pathologists were not the right sort of people to undertake forensic post-mortems, as they did not have sufficient forensic experience and the significance of some findings was beyond their skills and experience. Sometimes it only became clear that the death was suspicious, and therefore a forensic matter, when the post-mortem was undertaken. The time often came when the pathologist could do no more 'and science has taken over and provided the proof by chemical means'.[80]

Toxicology was a prime example of a field that was, by then, in the hands of forensic scientists in terms of detection of materials and estimation of quantities. The distribution of poison in the body also helped to establish the total amount consumed and helped give an estimate of time factors.[81] Indeed, Firth recounted an example of a poisoning case where the post-mortem had not revealed suspicious substances and where subsequent tests on tablets in the forensic science laboratory revealed that a fatal dose of morphine had been consumed.[82]

So the situation with regard to forensic pathology was unsatisfactory in the early years of Firth's career as a forensic scientist, a career which spanned the twenty years from just before the war, from the same time as the Home Office forensic science laboratory network had been in operation. Much of the problem with pathologists was due to a lack of available training in forensic medicine. When the regional forensic science laboratory network was established with the idea of having specialisms attached to the various laboratories so that there would be minimal overlap, only the Midlands laboratory had a pathologist as Director (Webster) as it was assumed that post-mortems would be done by local pathologists with Webster called in for difficult cases and for consultation.[83] However, this had not happened to the extent that was expected. So the situation at the end of the war was one of increasing specialization in forensic expertise, but without necessarily having the resources and organization to match it.

Understandably, the war years acted as something of a brake for the new forensic sciences. The demands of war put pressure on the medical services, and pathologists were required to use their skills in the identification of bomb victims.[84] Amongst the turmoil of war, the possibility of murders going unrecognized, being wrongly classified as the results of trauma or blast from bombs, was a considerable concern and therefore a threat to the maintenance of civil order in the UK. One such case was the Dobkin murder where pathologist Keith Simpson's brilliant scientific reconstruction of body parts with minimal remaining tissue, found in a bombed church in 1942, resulted in a conviction.[85] He was convinced that the remains belonged to a murder victim rather than a bomb victim and was able to link his findings to a missing woman, Rachel Dobkin.[86] A photograph of the skull taken by the Guy's Hospital photographer was fitted to a photograph of Rachel Dobkin, a technique which had first been used to dramatic effect in the Ruxton murder case.[87] There was sufficient tissue surviving round the throat for Simpson to establish that she had been strangled. Possibly, had the murder not taken place in wartime it would have become as well known in British forensic memory as the Crippen or Ruxton case.[88] Certainly Simpson in his autobiography suggested that 'it would have hit the headlines as Crippen did in Spilsbury's younger days'.[89] Interestingly, despite his central role, Simpson was not given top billing by the press. Guy's Hospital photographer Mary Newman was hailed as the star of the piece.[90] Indeed, Simpson had to write a tactful letter to the Metropolitan Police explaining why he did not use their photographic lab, and the police tended to cast him in a minor role compared with the main police investigation.[91] Nevertheless, the case was an example of teamwork across a broad spectrum of specialists: police; pathologist; the Home Office Analyst John Ryffel, who established that the victim had been buried in slaked lime which had a preservative effect; the dentist called in by Simpson to identify the victim's teeth and the photographer.[92] Through this case Simpson did much to demonstrate the importance of forensic pathology and its role in presenting scientific evidence, working in tandem with the police, at a time when the pressures of war and the possibility of covert murder threatened British society.[93]

Autobiography and forensic careers

Those who took up posts in the new forensic science laboratories were often 'headhunted', as in Firth's case where he was sought out because of his expert witnessing experience, or were proposed by someone in an appropriately senior position; witness Arthur Dixon accepting Sydney Smith's recommendation of his colleague, Dr James Davidson, for the post of Director of the Metropolitan Police Laboratory.[94] Candidates for the new forensic laboratory posts were viewed as potentially desirable appointments, not just because of their academic attainments, but also because of their experience in applying science to criminal and legal matters. For some scientists, involvement in criminal work and forensic experience happened more or less by accident. Other than a university degree which offered no training in forensic science, the only training in the discipline before the establishment of the forensic science laboratories was that available to public analysts or to the scientists at the Department of the Government Chemist, namely a two-year course in analytical chemistry at the Royal College of Science in London.[95] However, given that public analysts were trained in chemistry and sometimes medicine, a career as a public analyst was not an obvious route for physicists or mainstream biologists looking to gain experience of forensic work. At the beginning of the 1930s, then, when the establishment of state forensic science laboratories was becoming a reality, there was no available training in forensic science. Given the very limited pool of people suitable for the new forensic science laboratory roles, it is small wonder that great efforts were made to identify and recruit qualified and experienced candidates.

Several of the new forensic scientists had university connections rather than public analyst roots: Firth, the first Director of the North Western Laboratory, had come from a university appointment at Nottingham.[96] Tryhorn was Professor of Chemistry at Hull and heavily involved in forensic work and detective training before being appointed Director to the Harrogate Laboratory and then Home Office Forensic Science Adviser after the war.[97] However, forensic science only became a university subject in the latter part of the twentieth century, so the positions of the first forensic scientists, both in career and epistemological terms, were different from those working in forensic medicine. Whatever the deficiencies of forensic medical education, formal education in the subject was available from the end of the eighteenth century at the University of Edinburgh and from the 1830s at the University of Glasgow and in England. From 1831 the Society of Apothecaries required London medical students to take a three-month course in medical jurisprudence to qualify for its licence.[98]

Although published materials offer some clues to the careers of the new forensic scientists, biographical material is not plentiful and is largely confined to obituaries. However, a number of forensic scientists and medico-legal experts did produce autobiographies or memoirs, and some had biographies written about them; these provide a potentially rich source of insights into career and other aspects of the scientists' lives. Historians vary as to the level of veracity they are willing to credit to biography and autobiography. A.J.P. Taylor's oft-quoted comments that memoirs are 'set down to mislead historians' and 'useless except for atmosphere' serve as a warning.[99] Although, as Wallach notes, there is a commonly held view that autobiography cannot be objective, nevertheless historians of biography argue that biography or memoir may offer a potent way into understanding social and historical milieu which cannot readily be extracted elsewhere.[100] In addition the rise of 'life writing' celebrates subjectivity as a virtue, suggesting that, whatever their drawbacks, biographical texts can offer a rich picture of a life not available from other sources.[101] So, biographies, especially autobiographies and, here, memoirs, given that the focus of forensic biographies is on working lives rather than a whole life story, are potentially a significant resource.[102]

Biography can be especially useful for cataloguing scientific lives. Söderqvist argues that scientific biography is the most popular way that we read about science's past and it is now enjoying something of a renaissance, having fallen out of favour, to some extent, in the 1970s when history of science was enjoying its 'social turn'.[103] Although Söderqvist's arguments are made in relation to historians' practice of writing scholarly biographies of scientists, nevertheless, his points are also relevant in relation to scientists' popular biographies and autobiographies. In particular, he notes the joy of working on the scientist who lived a 'biographical life', keeping everything, including laboratory reports, correspondence, even down to the apparently trivial – bus tickets, shopping lists and the like – in the expectation that a biographer would one day find the material useful.[104] Such archives will yield rich pickings even if they threaten to overwhelm the biographer.

But most of these forensic scientists in this study did not lead a 'biographical life' and left no personal archive collections, no series of letters and no shopping lists, or if they did leave them, they did not put them where we can easily find them. However, Hamish Walls and J. B. Firth, senior figures in the new Home Office network of forensic science laboratories, did produce autobiographies.[105] In addition, there are two readily available 'colonial' biographies. John Thompson's autobiography, *Crime Scientist*, describes his work as Director of the Police Forensic Science Laboratory in Rhodesia (Zimbabwe) from 1963 to 1977, hence outwith the time frame of this book.[106] *Crime Chemist: The Life Story of Charles Anthony Taylor Scientist for the Crown* wears its colonial heart on its (book) sleeve, as Taylor worked in Australia.[107] These two scientists are not part of the story in the UK; nevertheless their working lives and entry into forensic work mirror that of British forensic scientists. Thompson claimed to have 'stumbled into a career in forensic science almost by accident', and the ways in which their work is presented through their biographies is similar to the styles of Walls and Firth.[108] It is also worth noting the memoirs of the internationally known and influential Swedish forensic scientist/criminologist (his memoirs describe him as a criminologist) and Director of the National Institute for Technical Police in Sweden, Harry Söderman which were written in English and published with a UK publisher (Longmans Green) and clearly intended to reach an Anglophone audience.[109] John McCafferty's memoir of his career as Cyril Cuthbert's assistant after the war, later taking charge of forensic ballistics in the Metropolitan Police Laboratory, is not a biography of a forensic scientist as such, but it provides insights into the Metropolitan Police Laboratory, its scientific staff and the shortage of resources after the war.[110]

The lives of medico-legal experts are better represented in biographical form with the published autobiographies of Glaister Jnr., Smith, Camps and Simpson.[111] Glaister and Smith led lives which were more biographical than those of their contemporaries in forensic science. Of course, such a small sample precludes generalization about family backgrounds; nevertheless the contrast between the affluent background of the Glaister dynasty, John Glaister Jr (1892–1971), son of John Glaister Snr (1856–1932), also Professor of Forensic Medicine at Glasgow, with talk of a big house, servants, motor car and financial security, stands in contrast to Firth's early education at night school and, through university scholarships, first steps on the path towards his Directorship of the North Western Forensic Science Laboratory.[112]

As Duvall demonstrates through his archive research, the autobiographies of the eminent medico-legists Glaister and Smith were ghost-written, partly because, at least in Smith's case, his initial draft was somewhat technical.[113] Given that the use of ghost writers is not usually public information, and given the lack of archive sources for the forensic scientists who were not as well known, this means that we do not know how much external editing or ghost writing was involved in the production of their biographies. However, even if explicit statements of concern about image and the use of ghost writers are not to be had, it is reasonable to infer that forensic scientists were just as concerned with self-image, or, if not self-image as such, then the image of their professional work and its status.

In the case of Firth, a concern to present the certainties of his professional role was indicated in his autobiography, where he presented scientific evidence as more conclusive in criminal cases than contemporary reports suggested. Ambage argues that there are some inaccuracies in Firth's account where cases were presented in a 'neat and tidy' way at odds with press reports of the same case, to the extent that there were sometimes 'quite startling discrepancies' between contemporary press reports and Firth's account.[114] For instance, in relation to the verdict in a rape and murder case in 1940, which Firth presented as a triumph of scientific analysis in his autobiography, contemporary press reports suggested the evidence was much less clear-cut; forensic expert witnesses disagreed about the link between forensic evidence and the accused.[115] The accused man was found guilty largely because of his confession, given that the forensic evidence presented by the North Western Laboratory was challenged by the evidence of the pathologist Webster, who was Director of the Birmingham Forensic Science Laboratory and had carried out the post-mortem. The evidence was also challenged by the expert witness for the defence, Professor W. H. Roberts, Liverpool's Public Analyst, who had often been used as an expert witness by the police before the Preston laboratory was opened. Science had not proved the case unequivocally, as Firth suggested; rather, the scientific evidence was ambiguous and the case itself presented the kind of disagreeing scientists scenario which Dixon was anxious to avoid, as he knew the damage courtroom discord could cause to the image of the new forensic science and the certainty and authority that it promised.[116]

Judging books by their covers (and their titles)

Presenting science as objective and decisive in criminal cases is one way in which biographies fostered a positive image for forensic scientists or medico-legal experts; photographs or other visual images are another. Most of these biographies (both medical and scientific) contained a number of black-and-white photographic images, unfortunately not always of significant quality and clarity. In some books there are graphic pictures relating to murders, but the abiding image tended to be a photo of the scientist, a middle-aged gent, wearing a well-used suit or heavy overcoat and fedora (attire akin to the Dick Tracy detective look, which was admittedly not uncommon for men in the UK in the 1930s and 1940s), bending rather unflatteringly over some (thankfully) rather indeterminate crime scene. This showed the scientist at work at the site of the crime, which, although possibly posed, was not a studio-quality photograph. It is not entirely clear that such a picture was particularly effective in enhancing the projected image of the scientist, although one has to be careful not to read the semiotics of the image through modern eyes. Given that 'scientific' crime scene management was a relatively new art, it is unlikely that there would be an immediately recognizable picture which would signify a scientifically managed crime scene in the cultural imaginary in the way that the modern crime scene, taped and sealed off with the scene of crime officer (SOCO) in a white protective suit is immediately recognizable to us today. The question was, then, what image would present the scientist as appropriately authoritative and scientific while at the same time connecting him to the arts of the new forensic sciences?

Cover images, even if photographs are reproduced in black and white, which are larger and of better quality are usually a better way of conveying an appropriate self-image.[117] Some forensic biography covers used abstract images, but where there was a drawing or photo of the medical man or scientist, it was overwhelmingly one of 'man with microscope'. The image of a microscope is, of course, something which can immediately be read as scientific, 'the iconic instrument of laboratory epistemology'.[118] Coupled with a white lab coat, this indicates that the wearer is an expert scientist. The dustcovers of the biographies of Firth, Glaister, Taylor, Thompson and Walls all displayed a photo or drawing of the scientist or medical man looking down his microscope.[119] Glaister, Taylor and Walls wore white coats; Firth and Thompson did not. Men looking through microscopes do not necessarily make aesthetically pleasing images; the face is often at least partly obscured. It is difficult to make out the detail of the microscope in a small photograph, and the cover of Walls's autobiography is an exemplar of why a small photograph of a man looking down a microscope does not necessarily make a good, semiotically clear image of a forensic scientist. On this cover, the scientist was looking down a *comparison* microscope, but you would not know this unless you were *au fait* with historical images of such instruments or unless you opened the book where the image was explained.[120] It is difficult to know what the instrument is from a casual glance at the cover; indeed, the plate containing the bullets for comparison has been clipped out of the image. It is interesting that the second edition of Walls's autobiography has a purely abstract cover with drawings of test tubes, unrealistic blood spatters and a kind of spring. The significance of the latter is unclear; perhaps it is meant to look like a particle track in a cloud chamber and therefore denoted an abstract, yet recognizable scientific image.[121]

In the same vein as 'man with microscope', Sydney Smith on the cover of his *Mostly Murder* arguably presented a more successful image.[122] This was a face-on, close-up head and shoulders image of a serious-looking, white-coated Smith of clearly advancing years, holding up and intently studying a test tube and pipette with a strap line 'Was he greater than Spilsbury?' Smith's criticism of Spilsbury was noted in Chapter 1, so his allusion to Spilsbury was pointed; other eminent forensic pathologists agreed with Smith's assessment.[123] The 'man with microscope' line drawing on Firth's book was accompanied by his name with all his initials denoting qualifications and professional membership and 'Director of the North Western Forensic Science Laboratory 1938–1958', presumably because he was not sufficiently well known for his readers to know who he was without being told.

The use of such images is not surprising, as these were clear ways of indicating masculine, scientific and epistemic authority, so this was a conscious effort to signify scientific status on the part of the medical and scientific forensic experts. Whereas the medical autobiographies could readily allude to murder in their titles (e.g., *Mostly Murder*, *Forty Years of Murder, Final Diagnosis*), forensic scientists tended to put crime in theirs (e.g., *A Scientist Turns to Crime, Crime Chemist*).[124] Söderman's (1957) memoirs are titled *Policeman's Lot: A Criminologist's Gallery of Friends and Felons*. The latter's cover photograph appeared at first sight to be a close-up view of handcuffed hands, a recognizable criminological image although not necessarily a scientific one; however, closer inspection reveals the hands appeared to have a double layer of chains around them rather than handcuffs, and a few centimetres of the arms were visible, revealing the sleeves of a lounge suit. The image, then, was distinctly ambiguous. Was this the criminal in chains or was it the 'policeman' whose lot may or may not have been a happy one?[125]

Further biographical details are to be found in Sweden's National Laboratory of Forensic Science's biographical pamphlet on Söderman which portrayed him with a more glamorous image than his British counterparts.[126] The cover image showed Söderman quizzically looking over his glasses while smoking a cigarette in a holder – a bloody handprint in red is superimposed on the black-and-white photo. Söderman's nickname was 'Revolver-Harry'.[127] By contrast, the cover of forensic pathologist Francis Camps's autobiography showed him smoking a pipe, and the back cover of Walls's biography, although not showing him smoking, portrayed him at his desk with the paraphernalia of pipe smoking clearly visible.[128]

Of course, smoking has long had glamorous connotations, and a cigarette holder was especially stylish, redolent of Hollywood, although not especially linked to masculinity in contrast to the tweedier, yet more definitively masculine image of pipe smoking.[129] Indeed Hilton notes the association of smoking, and especially pipe smoking, with late Victorian bourgeois masculinity.[130] As he points out, smoking figured large in the Sherlock Holmes canon.[131] This is not just in terms of Holmes's famous allusion to his elusive monograph on 140 different types of tobacco ash; it can also be seen in the surprisingly large number of references to smoking in the Holmes stories, where smoking is related to individuality.[132] Holmes literally drew upon his pipe when listening to the facts, using his pipe as an aid to working out the problem – the Red-Headed League was a 'three pipe problem' – so pipe smoking was linked to masculinity but was also linked to masculine intellect.[133] The stereotypical image of Holmes portrays him with deerstalker, pipe and magnify glass. It may be that Camps and Walls, in using pipe-smoking images, were acknowledging the power of the pipe in solving crimes à la Holmes.

The idea of smoking in a laboratory, especially a forensic laboratory, may appear somewhat unusual to modern sensibilities. Nevertheless, John McCafferty, when he was appointed as deputy to Cyril Cuthbert at the Metropolitan Police Laboratory after the war, recounted his introduction to the laboratory staff in his autobiography and was particularly struck with a pipe-smoking forensic scientist, Ian Holden would later become famous for his role in providing forensic expertise in the Great Train Robbery case in the 1960s.[134] Holden, who operated a spectrograph, had his teeth clamped round a large unlit S-shaped drooping pipe (the stereotypical Sherlock Holmes pipe no less) with a cork in the bowl to prevent the tobacco from getting wet and at the same time to stop the contents from falling on forensic evidence![135]

Although it would be tempting to argue that forensic scientists were more in need of visible displays of scientific status than their counterparts in forensic medicine, it is notable that Glaister and Smith both used scientific images on the dustcovers of their autobiographies. It is interesting that the projected self-image as a scientist appeared to be more important than displaying an image depicting some aspect of crime, but it was not necessarily easy to connect the two. Other than the question of displaying an image which might be just too gruesome for a dustcover, it is difficult to know what image would be instantly understood as a scientifically investigated crime scene. As already emphasized, we are used to modern readings of the semiotics of a crime scene cordoned off with crime scene tapes and with white-suited SOCOs moving efficiently about. If an image of a crime had been used as a cover image for Glaister's or Smith's autobiography, then it is difficult to see how it could have conveyed the scientific aspects separately from the police aspects of the crime. If an image was too criminological and less scientific in scope, there was a danger that a potential audience might have regarded the book as the memoir of a detective rather than a scientific or medical man, which could well have been viewed as diminishing status, given that a professor of forensic medicine would not have wanted to be mistaken for a detective, and could have lessened the perceived novelty of the book, given that many detective memoirs had been published. As an example of a cover which tried to signify science and crime, but not very successfully, the image on the cover of Australian forensic scientist Taylor's biography combined a man with microscope image in the top two-thirds of the page with a photo of two men in large coats bending over some disturbed earth.[136] It is not clear from the image whether they are scientists, police officers or someone else, and it is not obviously a crime scene.

Biography without autobiography (or biography)

Autobiography can be the product of retirement, so if an individual does not retire or only retires because of failing heath, then it is less likely that pen will be put to paper. Unfortunately, for some forensic scientists with fascinating careers, biographical details are much more elusive; this was often the case for independent scientific witnesses who often did not retire until they had to do so. Two of the best-known independent scientific expert witnesses of the first part of the twentieth century, although of different generations, were Charles Ainsworth Mitchell and Julius Grant. Neither of these two scientists produced an autobiography, although they were both prolific authors and enthusiastic office bearers in their professional societies (hence, presumably, interested in maintaining and promoting their professions beyond the business of making a living). Their professional society involvement centred on the Medico-Legal Society, Society of Public Analysts and, later, for Julius Grant, the Forensic Science Society. At various times,

Mitchell edited the Society of Public Analysts's journal, *The Analyst*, also jointly edited the *Medico-Legal Journal* and was the first chemist to be President of the Medico-Legal Society (Grant was the second, in 1973.)[137] As his forensic work expanded, Mitchell became known for his work on documents and inks, and he wrote several semi-popular works on science and crime.[138] A book cover image worthy of note, as it is a visual depiction of a forensic scientist in a line drawing, albeit not strictly belonging to a biographical work, can be found on the cover of Mitchell's popular book *The Scientific Detective and the Expert Witness*. The bespectacled scientist is depicted seated, wearing what appears to be a lab coat, but if so, not a white lab coat, next to a diagrammatic section of a human head; the scientist is measuring a human skull using a pair of dividers (other measuring tools – a ruler and compasses – are visible).[139] This image speaks of Lombrosian skull measurements rather than a modern forensic chemist at work in his laboratory!

Grant produced one semi-popular work, *Science for the Prosecution*, and co-wrote a textbook on fluorescence and ultra-violet analysis, with a substantial part on forensic applications of ultra-violet light; the latter ran to four editions between 1935 and 1954.[140] He also produced many publications on paper and papermaking technology, as his professional career centred on the paper industry. So it was not necessarily the case that these scientists had no time to write, let alone write an autobiography. Of course, producing an autobiography is not to everyone's inclination. The first three incumbents of the Directorship of the Metropolitan Police Laboratory, Davidson, Holden and Nickolls, did not do so, although, on the face of it, they all had time after they retired.[141] Cuthbert, also of the Metropolitan Laboratory, in his book, *Science and the Detection of Crime*, was keen to distance himself from the suggestion that his book was 'yet another autobiography'.[142] Perhaps as a police officer himself, his sensibilities were heightened by the number of detective memoirs on the market. Fortunately, his deputy John McCafferty had no such qualms about producing an autobiography wherein details of the parlous state of the Metropolitan Police Laboratory, post war, were revealed.[143]

For scientists relying on fees rather than a salary, where there was no prospect of an occupational pension, there was considerable impetus to continue working, possibly into their seventies and even eighties. This also meant that any book publication would have to be sufficiently lucrative a proposition so as to justify time spent away from paid expert witness work, given that they were also heavily involved in professional society work and editing. Bernard Spilsbury, operating on a fee basis, never had time to write his promised textbook on forensic medicine, let alone an autobiography, and there was speculation that his suicide in his laboratory at the age of seventy reflected his failing health and the fact that he could not retire, as he relied on fee income.[144] It is likely that Firth and Walls would have been in receipt of Civil Service pensions when they took up their pens even though they were still active in their disciplines. Although Mitchell and Grant both had long-running salaried positions (British Vinegar Company and John Dickson Paper Company, respectively), we do not know whether they were as fortunately superannuated as Firth and Walls were likely to have been. In any case, they may well have enjoyed their professional activities and wanted to continue them as long as possible. Grant's two pieces, 'Forensic scientist at large' and 'Forensic scientist still at large', were not the words of a man who was tired of his discipline and thinking of retirement![145] Indeed, Mitchell only retired on health grounds in 1945 and was still active in the Medico-Legal Society until shortly before his death.[146] Grant was famously consulted as to the authenticity of the notorious Hitler diaries in the 1980s and other matters when he was well into his eighties.[147] Neither of them seemed to have much of a retirement in which to contemplate writing an autobiography.

Forensic memoirs as detective memoirs

Given the caveats about biography and the concerns of forensic scientists and medico-legal experts to present an image which was appropriately mature, knowledgeable, high status and scientific, if sometimes ambiguous, in criminal investigation, how, then, should we view these biographies and memoirs and, importantly, how should they be read as part of an existing genre? They were certainly not heirs to the Victorian 'life and letters' style of scientific biographies that were written to reflect great men and great ideas (usually biographies written after the great man's death).[148] This type of biography had more or less disappeared by the middle of the twentieth century when the medico-legal experts and first generation of forensic scientists were turning out their autobiographies, books that were altogether aimed at a more popular audience with an appetite for 'true crime' stories. Indeed, 'Working Life and Crimes' might be a more appropriate designation for these works, with the emphasis on the latter. This suggests that it is much more fruitful to think of them as variants of the police or detective memoir.

There are aspects of works in the detective memoir genre which are very similar to these 'forensic memoirs'. Although some detective memoirs were fictional, many were written by retired detectives who described cases, even if they were leaving out the boring bits. From the late nineteenth century and well into the twentieth, detectives were far more likely than other occupational groups to write memoirs and so there are many of these works.[149] Given that publishers did not usually publish memoirs of working-class people and most detectives came from working-class backgrounds, having risen through the ranks of the police force to become detectives, these were unusual literary texts. But given a better-educated public from the end of the nineteenth century with an appetite for reading, given an enthusiasm for crime literature (the more sensational, the better), and given the boost to such interest from the hugely appealing Sherlock Holmes from 1887 onwards, it is not surprising that detective memoirs should have done well. The story of the detective was attractive as he appeared to offer certainty, pursuing truth, solving crimes and restoring order at a time when other certainties about social and economic life were being shaken.[150]

As Shpayer-Makov notes, detective memoirs followed a formula: a short introduction and then a much longer section of work experience presented the 'occupational self'; they were work histories rather than life histories.[151] Given that the police were subject to plenty of criticism in real life and fictional private detectives often portrayed the police as ineffective or bumbling, this resulted in 'galvanizing the detective-memoirists into action, mounting a counter-offensive against detective fiction. Repeated disparaging references to this literature in their memoirs gave their books a kind of crusading aura in battling the misrepresentation of police detectives'.[152] Detective memoirists were particularly keen to emphasize that fictional detectives' methods would not work, emphasizing the contrast with their real-life jobs which were difficult, not to say gruelling, with hours spent tracking down the criminal, 'unlike the private middle-class detective who often solved puzzling crimes while sitting in his armchair and using scientific deduction'.[153] The positive message of the detective memoir may have hit its mark as the image of English detectives began to improve after the First World War despite the negative ways in which they had often been portrayed in detective fiction.[154]

Although the biographies of forensic scientists and medico-legal specialists were not detective memoirs, no matter how much they drew inspiration from the success of the latter, there is at least one detective memoir from a police officer who knew the workings of an early forensic laboratory in detail. This was John McCafferty's *Mac, I've Got a Murder*.[155] As a serving police officer, he was appointed as a liaison officer, Cyril Cuthbert's assistant, at the Metropolitan Police Laboratory after the war as the lab began to find its feet again. McCafferty confirmed the shoddy treatment of Cuthbert before the lab opened in the 1930s, when the latter was almost fired over his involvement with forensic work.[156]

Hence, there is every reason to suggest that forensic memoirs were written to conform with, and to capitalize on, the considerable success of the detective memoir. Forensic autobiographies tended to start with a chapter on early life, showing how the author moved into forensic work, followed by chapters describing interesting and important cases, in effect, a work history. Forensic scientists and medico-legists often nodded to Sherlock Holmes as an influence (sometimes covertly as in the 'pipe' image) – allusions to Holmes popped up everywhere.[157] With varying degrees of assertiveness, they, too, pointed out the ways in which Sherlock Holmes's methods would not have been up to the mark in real life. Although not part of his autobiography, Keith Simpson, Home Office Pathologist from the 1930s, although clearly a fan of Sherlock Holmes to the extent that he delivered a lecture to the Sherlock Holmes Society in 1953, was scathing about Holmes's scientific analyses, or lack of them, in a much later pamphlet, and even took poor old Dr Watson to task over his medical knowledge and professional attitude.[158]

On the first page of his autobiography, indeed in the very first paragraph, Firth invoked the memory of Holmes, Maigret and Poirot. He went on to discuss the influence of Sherlock Holmes:

> Holmes embodied characteristics one would expect to find in a whole group of forensic scientists at one of our major laboratories. Holmes, however, had one great advantage over the police officer and the forensic scientist alike; he did not have to go into the law courts and expose his evidence to impartial judgment.[159]

Conclusion

The roots of forensic science expertise are discernible through the Home Office Analysts and the system of public analysts, where much expertise in forensic chemistry was developed; the latter role was influential in the careers of a number of prominent independent forensic consultants, before posts in the new forensic laboratories became available, for what was usually an older generation of scientists involved in forensic analysis. Long dynasties of master and apprentice prevailed in analytical chemistry. Other roles involving forensic analyses were to be found in government laboratories in the UK and in her colonies.

The autobiographies and biographies described in this chapter are a form of literary production, and it is to a different form of literary production that the final chapter turns. Given the immense popularity of detective fiction from the last part of the nineteenth century onwards, given the widespread enthusiasm for the scientific detective amongst the public, who were also consuming newspaper reports on dramatic forensic cases, and given that forensic scientists often alluded to scientific detectives in their popular writings, a consideration of how scientific detective fiction relates to the cultural imaginary of forensic science is the last piece of the jigsaw.

Notes

1 Julius Grant, as a paper and ink expert, achieved considerable public acclaim for his debunking of the Hitler diaries in the 1980s when he was well into his eighties. See D. C. Chamberlain, 'Paper', in M. F. Suarez, and H. R. Woudhuysen (eds), *The Book: A Global History*, Oxford: Oxford University Press, 16–29, p. 128.
2 H. J. Walls, *Expert Witness: My Thirty Years in Forensic Science*, London: John Long, 1972, p. 15.
3 K. D. Watson, 'Medical and chemical expertise in English trials for criminal poisoning, 1750–1914', *Medical History*, 2006, 50: 373–390, 374.
4 K. D. Watson, *Forensic Medicine in Western Society: A History*, Abingdon: Routledge, 2011, pp. 57–58.
5 K. D. Watson, 'Medical and chemical expertise', p. 386.
6 Ibid., p. 375.
7 J. Ward, *Origins and Development of Forensic Medicine and Forensic Science in England, 1823–1946*, unpublished PhD thesis, The Open University, 1993, p.78.
8 Ibid, p. 84.
9 Ibid., p. 4.
10 K. D. Watson, *Poisoned Lives: English Poisoners and Their Victims*, London and New York: Hambledon and London, 2004, p. 172.
11 Ward, *Origins*, p. 4.
12 Ibid., p. 6.
13 Ibid., p. 84.
14 Ibid., p. 86.
15 Ibid., p. 82.
16 N. Ambage, *The Origins and Development of the Home Office Forensic Science Service, 1931–1967*, unpublished PhD thesis, Lancaster University, 1987, p. 70.

17 Ward, *Origins*, p. 236.
18 Ibid.
19 G. Taylor, *Forensic Enforcement: The Role of the Public Analyst*, Cambridge: RSC Publishing, 2010, p. 7.
20 P. W. Hammond and H. Egan, *Weighed in the Balance: A History of the Laboratory of the Government Chemist*, London: HMSO, 1992, p. 87.
21 B. Dyer and C. A. Mitchell, *The Society of Public Analysts and Other Analytical Chemists: Some Reminiscences of Its First Fifty Years and Review of Its Activities*, Cambridge: Heffer, 1932.
22 P. Atkins, *Liquid Materialities: A History of Milk, Science and the Law*, Farnham: Ashgate, 2010, p. 107.
23 Watson, *Poisoned Lives*, p. 171.
24 Watson, 'Medical and chemical expertise', p. 388.
25 Watson, *Poisoned Lives*, p. 172.
26 Ambage, *The Origins*, p. 131.
27 Ibid.
28 Hammond and Egan, *Weighed in the Balance*, p. viii.
29 Ibid. p. 33. A. Stanziani, and P. J. Atkins, 'From laboratory expertise to litigation: The Municipal Laboratory of Paris and the Inland Revenue Laboratory in London, 1870–1914: A comparative analysis', in C. Rabier (ed), *Fields of Expertise: A Comparative History of Expert Procedures in Paris and London, 1600 to Present*, Newcastle upon Tyne: Cambridge Scholars Publishing, 2007, 317–338, p. 327.
30 Hammond and Egan, *Weighed in the Balance*, p. 88.
31 Ibid., p. 90.
32 R. C. Chirnside and J. H. Hammence, *The 'Practising Chemists' A History of the Society for Analytical Chemistry 1874–1974*, London: The Society for Analytical Chemistry, 1974, p. 69.
33 J. Steere-Williams, 'A conflict of analysis: Analytical chemistry and milk adulteration in Victorian Britain', *Ambix*, 2014, 61 (3): 279–298, p. 283.
34 Ibid., p. 287.
35 Hammond and Egan, *Weighed in the Balance*, p. 156.
36 Steere-Williams, 'A conflict of analysis', p. 279.
37 Hammond and Egan, *Weighed in the Balance*, p. 187.
38 E.g., C. T. Symons. 'Scientific aids in prospect', *Metropolitan Police College Journal*, 1935, 1 (2): 30–34.
39 M. Gilberg, 'Alfred Lucas: Egypt's Sherlock Holmes', *Journal of the American Institute for Conservation*, 1997, 36 (1): 31–48.
40 A. Lucas, *Ministry of Finance, Egypt, Report on the Work of the Government Analytical Laboratory and Assay Office During the Period 1913–19 by A. Lucas, F.I.C., Director*, Cairo: Government Press. 1920, p. 1.
41 Ibid., p. 8.
42 Ibid.
43 Ibid. Such cases were presumably sent to the medico-legal expert.
44 Ibid.
45 Ibid.
46 Ibid.
47 Lucas stated this in the 1939 broadcast: 'Tutankhamun's trumpets', BBC CDA21502. 16 April 1939. See C. Finn, 'Recreating the sound of Tutankhamun's trumpet', available at http://www.bbc.co.uk/news/world-middle-east-13092827, accessed 11 May 2015.

48 A. Lucas, 'Government Analytical Laboratory, Cairo, Report of the Director for the Year 1920', D. R. Wood and R. B. Pilcher, 'Notes from the Reports of Public Analysts', *The Analyst*, 1922, 47: 19–24, p. 22.
49 Gilberg, 'Alfred Lucas'.
50 Hammond and Egan, *Weighed in the Balance*, p. 203.
51 C. A. Mitchell, 'Obituary. Otto Hehner', *The Analyst*, 1924, 49 (584): 501–505, p. 501. For details on William Ramsay's career, see K. D. Watson, 'The chemist as expert: The consulting career of Sir William Ramsay', *Ambix*, 1995, 42 (3): 143–159.
52 Mitchell, 'Otto Hehner', p. 502.
53 Ibid., p. 505.
54 H. E. Cox, 'Charles Ainsworth Mitchell (1867–1948), *The Analyst*, 1948, 73: 55–57 and Mitchell, 'Otto Hehner'.
55 See Cox, 'Charles Ainsworth Mitchell' for a description of Mitchell's life, work and publications.
56 Cox, 'Charles Ainsworth Mitchell' and K. Macrakis, *Prisoners, Lovers, and Spies: The Story of Invisible Ink from Herodotus to al-Qaeda*, New Haven, CT and London: Yale University Press, 2014, p. 132.
57 Macrakis, *Prisoners, Lovers, and Spies*, p. 133.
58 S. Horrocks, *Consuming Science: Science, Technology and Food in Britain, 1870–1939*, unpublished PhD thesis, University of Manchester, 1993, pp.149–150.
59 Ibid.
60 E. Voelcker and R. S. Watson, 'Obituary: Henry Edward Cox', *The Analyst*, 1952, 77 (913): 169–170.
61 K. D. Watson, 'The chemist as expert', p. 151.
62 W. Jago, (1909) *A Manual of Forensic Chemistry Dealing Especially with Chemical Evidence, Its Preparation and Adduction. Based upon a Course of Lectures Delivered at University College, University of London.* London: Stevens and Haynes and, A. Lucas, *Forensic Chemistry*, London: Edward Arnold, 1921.
63 W. Jago, *The Chemistry of Wheat, Flour, Bread and Technology of Breadmaking*, Brighton: William Jago, 1886, and W. Jago, and W. C. Jago, *The Technology of Bread-Making: Including the Chemistry and Analytical and Practical Testing of Wheat, Flour, and Other Materials Employed in Bread-making and Confectionery*, London: Simpkin, Marshall, Hamilton, Kent & Co., 1911.
64 D. W. Kent-Jones, 'Obituary notices: William Jago 1853–1938' *Journal of the Chemical Society*, 1938, 1127–1128.
65 S. Smith, *Mostly Murder*, London: Harrap, 1959.
66 Ibid., p. 38.
67 Ibid.
68 Ibid., p. 98.
69 I am grateful to Heather Wolffram for pointing out evidence of their information sharing through Lucas's reports in Smith's *Royal College of Physicians of Edinburgh* archive.
70 Smith, *Mostly Murder*, p. 98.
71 Ibid., pp. 98–114.
72 A. Lucas, *Forensic Chemistry and Scientific Criminal Investigation*, London: Edward Arnold, 1945, 4th edn., pp. 21–22.
73 Smith, *Mostly Murder*, p. 35.
74 C. A. Mitchell, *Forensic Chemistry in the Criminal Courts*, London: The Institute of Chemistry of Great Britain and Northern Ireland, 1938, p. 3.
75 Ibid., p. 13.

76 Ibid., p. 17.
77 J. B. Firth, *A Scientist Turns to Crime*, London: William Kimber, 1960, p. 14.
78 Ibid., p. 15.
79 Ibid., p. 19.
80 Ibid., p. 20.
81 Ibid., p. 21.
82 Ibid., p. 172.
83 J. B. Firth, 'Forensic science laboratories', *Medico-Legal and Criminological Review*, 1945, 13 (3): 120–133, p. 126.
84 A. Bell, 'The development of forensic pathology in London, England: Keith Simpson and the Dobkin case, 1942', *Canadian Bulletin of Medical History/Bulletin Canadien D'histoire de la Médecine*, 2012, 29 (2): 265–282, p. 266.
85 Ibid., p. 266.
86 Ibid., p. 271.
87 See Smith, *Mostly Murder*, pp. 227–236, for a discussion of the notorious Ruxton case.
88 Bell, 'The development of forensic pathology', p. 272.
89 K. Simpson, *Forty Years of Murder: An Autobiography*, London: Harrap, 1978, p. 49.
90 Bell, 'The development of forensic pathology', p. 273.
91 Ibid., p. 274.
92 Simpson, *Forty Years of Murder*, pp. 50–52.
93 Bell, 'The development of forensic pathology', p. 275.
94 See Firth, *A Scientist* for J. B. Firth and Ambage, *The Origins*, for J. Davidson.
95 Hammond and Egan, *Weighed in the Balance*, p. 158.
96 Firth, *A Scientist*.
97 H.J. Walls, 'The Forensic Science Service in Great Britain: A short history', *Journal of the Forensic Science Society*, 1976, 16: 273–277, p. 274.
98 Watson, *Forensic Medicine*, pp. 57–58 and I. Burney, *Poison, Detection, and the Victorian Imagination*, Manchester: Manchester University Press, 2006, p. 43.
99 J. J. Wallach, 'Building a bridge of words: The literary autobiography as historical source material', *Biography*, 2006, 29 (3): 446–461, p. 447.
100 Ibid., p. 448 and K. V. Hansen, 'Historical sociology and the prism of biography: Lillian Wineman and the trade in Dakota beadwork, 1893–1929', *Qualitative Sociology*, 1999, 22 (4): 353–368.
101 S. Smith and J. Watson, *Reading Autobiography: A Guide for Interpreting Life Narratives*, Minneapolis: University of Minnesota Press, 2001.
102 J. D. Popkin, *History, Historians & Autobiography*, Chicago, IL and London: University of Chicago Press, 2005.
103 T. Söderqvist, 'What is the use of writing lives of recent scientists?', in R. E. Doel and T. Söderqvist (eds), *The Historiography of Contemporary Science, Technology and Medicine: Writing Recent Science*, Abingdon and New York: Routledge, 2006, 99–127, p. 99.
104 Ibid., p. 101.
105 Firth, *A Scientist*; Walls, *Expert Witness*.
106 J. Thompson, *Crime Scientist*, London: Harrap, 1980.
107 A. Dower, *Crime Chemist: The Life Story of Charles Anthony Taylor Scientist for the Crown*, London: J. Long, 1965.
108 Thompson, *Crime Scientist*; see book jacket.
109 H. Söderman, *Policeman's Lot: A Criminologists' Gallery of Friends and Felons*, London: Longmans, Green and Co., 1957.
110 J. McCafferty, *Mac, I've Got a Murder*, London: Arthur Barker Ltd., 1975.

111 J. Glaister Jnr., *Final Diagnosis*, London: Hutchinson, 1964; Smith, *Mostly Murder*; F.E. Camps, *Camps on Crime*, Newton Abbot: David & Charles, 1973; Simpson, *Forty Years of Murder*.
112 See Glaister, *Final Diagnosis* and Firth, *A Scientist*, Chapter one.
113 N. Duvall, *Forensic Medicine in Scotland, 1914–39*, unpublished PhD thesis, University of Manchester, 2013, pp. 218–219.
114 Ambage, *The Origins*, p. 188 and Firth, *A Scientist*.
115 Ambage, *The Origins*, p. 189.
116 Ibid., p. 190.
117 All the books cited in this section are still in copyright; therefore, images have not been included.
118 G. Gooday, 'Placing or replacing the laboratory in the history of science?', *Isis*, 2008, 99 (4): 783–795, p. 792.
119 Firth, *A Scientist*; J. Glaister, *Final Diagnosis*, London: Hutchinson, 1964; A. Dower, *Crime Chemist: The Life Story of Charles Anthony Taylor Scientist For the Crown*, London: J. Long, 1965; J. Thompson, *Crime Scientist*, London: Harrap, 1980; Walls, *Expert Witness*.
120 Walls, *Expert Witness*, Figure 2a (after p. 96).
121 H. J. Walls *Expert Witness: My Thirty Years in Forensic Science*, London: The Quality Book Club, 1973, 2nd edition.
122 Smith, *Mostly Murder*.
123 See Simpson, *Forty Years of Murder*. Simpson was highly critical of Bernard Spilsbury and his lack of interest in mentoring younger members of the forensic pathology profession. On the other hand, he spoke warmly of Sydney Smith and the advice and help he gave to less experienced colleagues.
124 See Smith, *Mostly Murder*; Simpson, *Forty Years of Murder;* Glaister, *Final Diagnosis* and Firth, *A Scientist*; Thompson, *Crime Chemist*.
125 It is interesting that the allusion to Gilbert and Sullivan (A policeman's lot is not a happy one) is a line from the song, 'A policeman's lot' from *The Pirates of Penzance* and is made by the title and repeated inside. This suggests that the book may have been intended for a UK audience who would be very familiar with such a cultural reference. See I. Bradley, *The Annotated Gilbert and Sullivan*, Harmondsworth: Penguin Books, 1982 for details of Gilbert and Sullivan operas.
126 I. Kopp, *The Fantastic Life of Harry Söderman 1902–1956*, National Laboratory of Forensic Science – SKL/National Forensic Centre. Available at http://nfc.polisen.se/Global/www%20och%20Intrapolis/Informationsmaterial/SKL/Soderman_minnesskrift.pdf, 2015, accessed 21 April 2015.
127 Ibid.
128 Camps, *Camps on Crime*; Walls, *Expert Witness*.
129 M. Pugh, *We Danced All Night: A Social History of Britain Between the Wars*, London: Vintage, 2009, pp. 222–223.
130 M. Hilton, *Smoking in British Popular Culture 1800–2000: Perfect Pleasures*, Manchester and New York: Manchester University Press, 2000.
131 Ibid., 17–20.
132 See A. Conan Doyle, 'Boscombe Valley Mystery', in *The Adventures of Sherlock Holmes*, 1892. Available at http://www.gutenberg.org/files/1661/1661-h/1661-h.htm for monograph on fourteen types of tobacco ash, accessed 28 April 2015.
133 Conan Doyle, 'The Red Headed League', in *The Adventures of Sherlock Holmes*, George Newnes Ltd, 1892. Available at http://www.gutenberg.org/files/1661/1661-h/1661-h.htm, accessed 28 April 2015.

134 See A. Scaplehorn, Obituary: Ian George Holden BSc PhD AKC FRIC, 1922–2011, *Science and Justice*, 2011, 51: 216. Ian Holden is not to be confused with H. S. Holden (no relation according to Cafferty, *Mac, I've Got a Murder*, p. 19) who took over as Director of the Metropolitan Laboratory in 1946 (Ibid., 26). Scaplehorn confirms that Ian Holden often plugged up his pipe with a cork and put it in his pocket. 'On one occasion he was in the witness box at the Old Bailey when the Judge leaned forward and commented on the smoke rising from Ian's jacket pocket.' Scaplehorn, 'Obituary'.
135 Cafferty, *Mac, I've Got a Murder*, pp. 17–18.
136 Dower, *Crime Chemist.*
137 Cox, 'Charles Ainsworth Mitchell'; J. Grant, 'Presidential address: Forensic scientist at large', *Medico-Legal Journal*, 1973, 41 (4): 132–141.
138 C. A. Mitchell, *Science and the Criminal*, London: Isaac Pitman & Sons, 1911; C. A. Mitchell, *The Scientific Detective and the Expert Witness*, Cambridge: W. Heffer & Sons Ltd, 1931; C. A. Mitchell, *A Scientist in the Criminal Courts*, London: Chapman & Hall Ltd, 1945.
139 Mitchell, *The Scientific Detective*, front cover.
140 J. Grant, *Science for the Prosecution*, London: Chapman and Hall, 1941; J. A. Radley, and J. Grant, *Fluorescence in Ultra-Violet Light*, London: Chapman and Hall, 1933 (4th edition 1954).
141 K. Simpson, 'Foreword', H. J. Walls, *Expert Witness*, 11–12, p. 11.
142 C.R.M. Cuthbert, *Science and the Detection of Crime*, London: Hutchinson, 1958, p. 9.
143 McCafferty, *Mac, I've Got a Murder.*
144 A. Rose, *Lethal Witness: Sir Bernard Spilsbury Honorary Pathologist*, Chalfont, Stroud: Sutton, 2007, p. 268.
145 Grant, 'Presidential address: Forensic scientist at large'; J. Grant, 'Forensic scientist – Still at large!' *Medico-Legal Journal*, June 1982, 50 (2): 61–74.
146 Cox, 'Charles Ainsworth Mitchell'.
147 Anon., 'Julius Grant 1901–1991', *Journal of the Forensic Science Society*, 1991, 31 (3): 397–399.
148 M. J. Nye, 'Scientific biography: History of science by another means?', *Isis*, 2006, 97 (2): 322–329.
149 H. Shpayer-Makov, 'Explaining the rise and success of detective memoirs in Britain', in C. Emsley and H. Shpayer-Makov (eds), *Police Detectives in History, 1750–1950*, Aldershot and Burlington, VT: Ashgate, 2006, pp. 103–133.
150 Ibid., p. 107.
151 Ibid., pp. 114–115.
152 Ibid., p. 121.
153 Ibid., p. 125.
154 Ibid., p. 132.
155 McCafferty, *Mac, I've Got a Murder.*
156 Ibid., p. 26.
157 For example, two such allusions include Firth, *A Scientist*, p. 13 and J. Grant, 'The past, present and future ro1e of the private forensic science laboratory', *Journal of the Forensic Science Society*, 1977, 16 (3): 197–200, p. 197.
158 K. Simpson, *Sherlock Holmes on Medicine and Science*, With an Introduction by Isaac Asimov BSI and An Appreciation by E. Stanley Palm, New York: Magico Magazine, 1983.
159 Firth, *A Scientist*, p. 13.

6 Forensic science and forensic fiction

Introduction

In order to understand the development of forensic science in Britain and the lives of forensic scientists, preceding chapters concentrated on traditional sources. This included a wealth of secondary material on criminology, policing, history of science and science and technology studies, newspaper accounts, articles and books written by the scientists and, of course, biographies and autobiographies. However, these sources cannot reveal all of the history of forensic science; some of the story is told through crime and detective fiction. It would be difficult to find another scientific discipline where literary characters had quite such an influence on the public imagination and on the scientists themselves.

Although the focus of this chapter is on detective fiction, this is set against the backdrop of a broader scientific fiction which reflected the fight for professional status which many scientists, of whatever stripe, experienced. Even if detective fiction offers little by way of realistic role models, the possibility that real (or nearly real) forensic sciences and technologies were anticipated by fictional scientific detectives is irresistibly fascinating to a wide audience ranging from critics of detective literature and literary-minded scientists to eager consumers of crime fiction.[1] If we consider the popularity of contemporary television forensic dramas such as *CSI*, we might imagine that the 'forensic turn' is a recent phenomenon.[2] But this was not so – previous generations enjoyed just as much of a 'forensic turn' as we do now. The development of the literature on scientific detection is as much a part of the 'forensic turn' as the development of real-life scientific detection, forensic science laboratories and forensic careers.

The 'forensic turn' does not end there. There has been a long-standing interest in connecting the real and the fictional in the forensic sciences, and an analysis of the putative connection, even if it sometimes appears tenuous, offers valuable clues about popular understanding of the history of forensic science. The postulated connection is more subtle and broader than a simple two-way influence between crime/detective fiction and forensic science and vice versa. Even describing such a relationship as two-way – fiction influences science/science influences fiction – makes the link seem much more one-dimensional than it was. Instead, connections between detective literature and forensic science are better viewed as a reflection of a more pervasive phenomenon, a common cultural imaginary, where science and the battle against crime became firmly interwoven in a shared set of concerns for establishing and maintaining social and moral order at the turn of the nineteenth century, a concern with order that is manifest right through the development of the forensic sciences.[3] 'Cultural imaginary', a wider term than forensic turn, captures the connection and the long-lasting nature of this relationship. Such a cultural imaginary embraces the urge to bring a broadly understood scientific approach to criminal detection literature, in the same spirit as activities that were badged as scientific in real-life detection.

Fictional detectives are not the only characters who have something to tell us about the development of forensic science. Indeed, whatever they might offer by way of actual scientific analyses (or not), they are somewhat tenuous role models for professional scientists. Fictional portrayals of scientists and scientific professions across the canon of scientifically based literature mirrored scientists' anxieties about status and control. We would hardly expect real-life scientists involved in forensic disciplines to be immune to such professional worries, given the novelty of many of their roles in forensic work. Their status was not exactly marginal – given the impetus to establish state forensic science laboratories in England and Wales in the 1930s, appropriate expertise was in some demand. Rather, their status was new, without the sedimented history of more established scientific and medical roles. As the preceding chapter shows, such concerns were expressed by forensic scientists in their own autobiographies and professional writings. Forensic scientists *did* write about status and professional demarcation, so when Mitchell criticized the Scientific Advisory Committee for sidelining forensic chemistry or Firth complained about the unsatisfactory state of forensic pathology in England, they were making statements about demarcation and status, concerns which can also be found in fictional lives.[4]

Forensic sciences in fiction and fiction cited by forensic scientists in their technical and popular writing reflect a wide-ranging belief that science and scientists could be successfully enrolled in the fight against crime – indeed, a belief that science was *fundamental* to the successful control of crime, an extension of the rhetoric that a scientific approach was vital for beating crime in real life. Hence, the rise of detective fiction and the development of forensic science can both be read as responses to perceived problems of crime, order, social control and the control of colonial subjects. Wider worries about control of social and moral order, not only control of crime, were reflected in the development of both the real-life sciences and the fiction. Such fears centred on crises of masculinity degeneracy of the race, eugenics, concern with criminal types and loss of empire.[5] Detective fiction and evolving forensic scientific practice were both imbricated in creating important aspects of culture, offering models of ways in which society might be ordered and controlled, where the period from the Edwardian era onwards is crucial. A subtle and pervasive cultural imaginary is involved in the mutual development of forensic fictions and forensic realities.

The perennial question of what counts as 'scientific' is also relevant in fiction. In fiction as in life, 'scientific' was often taken as a synonym for modern, progressive, ordered and controlled rather than attaching to actual scientific analyses or theories. One of the ways we construct, consume, define, perform, bring into being and think about science is through fiction. Van Dover, a perceptive commentator on the role of science in scientific detective fiction, notes that it is largely the enrolment of 'the scientific method' rather than the demonstration of scientific analyses in detective fiction that marks such fiction as scientific.[6] Hence, it is productive to see the development of detective fiction and the development of forensic science from the Edwardian period onwards woven into other fictional and non-fictional influences in the making of a culture where forensic science would develop and flourish.

Thomas argues that developments in forensic sciences and technologies had significant influence on detective fiction. Sherlock Holmes is universally known, but several other scientific detectives were created in his wake.[7] Although some achieved considerable popularity in their time, none has achieved the lasting popularity of Holmes. It would be possible to consider the vast array of fictional detectives who enrolled science in some shape or form in their detective work, but such a task would take up a whole book. Instead the focus is on Sherlock Holmes – such is his influence that he is impossible to ignore – but, specifically, Sherlock Holmes in comparison to Dr John Thorndyke, the fictional medico-legal expert devised by Richard Austin Freeman (1862–1943), who had received an education in medicine.[8] The appeal of Thorndyke is discussed in more detail later in the chapter, but he is introduced briefly here in order to make a comparison with Holmes. As the two most popular scientific detectives in the UK spanning the late nineteenth and first part of the twentieth century, their fictional careers demonstrate interestingly different aspects of the use of science in detective work.

Van Dover's study of scientific method and scientific literary detectives lists Sherlock Holmes, John Thorndyke and Craig Kennedy as the most popular fictional scientific detectives of the late nineteenth century (for Holmes) and first third of the twentieth century in the USA and UK (for all three).[9] Arthur B. Reeve wrote the Craig Kennedy stories between 1910 and the 1930s.[10] Kennedy was a fictional chemistry professor in Columbia University, and the action takes place in and around New York. The Kennedy stories are shot through with a variety of scientific and technical gizmos, technoscience *par excellence*. Craig Kennedy was particularly fond of elaborate surveillance technologies and he was not overly fussy about the ethical dimensions of such technologically aided snooping. It is tempting to suggest that he would have been quite at home in the early part of the twenty-first century. He treated psychoanalysis as an exact science. The Kennedy tales are far-fetched in comparison to the Thorndyke stories; characterization and plot are sacrificed to high-tech action. The stories have a distinctly modern feel to them, much more so than Thorndyke's adventures. Kennedy most definitely lives in the twentieth century. Nevertheless, these tales were immensely popular in their day in the UK and especially in the USA. It is testimony to readers' appetite for Holmes and scientific detection that scientific detectives of doubtful literary merit could be pulled along in his slipstream and enthusiastically embraced. Kennedy would be worthy of study for no other reason than his sheer popularity. However, he is not included in my discussion of literary detectives, partly for reasons of space but mainly because he did not work in the UK and, to my knowledge, he is never mentioned in the work of British forensic scientists. If he was popular in the UK, he does not seem to have entered the collective forensic memory of the British the way Holmes certainly did and Thorndyke did to a lesser, but still tangible, extent. There appear to be no current print editions of Kennedy in the UK, whereas the Thorndyke stories and novels have been regularly reprinted there.

One of the ways in which fictional influences on professional roles may be tracked is through the claims of those involved in forensic detection to have been influenced by fiction and, conversely, where those producing fiction have explicitly modelled their characters on existing forensic experts. Of course, we must recognize that a good deal of rhetoric is involved in explicit claims of influence. Indeed, it was so common for those involved in forensic science in the first half of the twentieth century to allude to the influence of Sherlock Holmes in biographical or general books or essays that a nod to Holmes became a recognized and almost obligatory rhetorical anchor for the emerging cadre of forensic scientists in their popular works and in introductions to books. However, as we shall see in what follows, not everyone was convinced of the scientific skill of Sherlock Holmes to provide a suitable role model.

An analysis of the careers and cases of fictional scientific detectives may usefully be set against prevailing stereotypes of scientists, or rather, 'men of science' in late Victorian fiction and in the fiction of the first half of the twentieth century. However, there is a caveat. Scientific detectives were often cast as detectives first and scientists second, no matter how far they were portrayed as expert scientists. Sherlock Holmes is often regarded as the archetype of this literary model. Although the opening scenes of the first Holmes novel, *A Study in Scarlet* (1887), introduced the detective experimenting in St Bartholomew Hospital's chemistry laboratory, Holmes later made it clear that his vocation was one of consulting detective; indeed, he regarded his calling as unique, he was the only consulting detective in the world.[11] Nevertheless, there were sufficient 'scientist first' literary detectives, with Dr John Thorndyke and Professor Craig Kennedy as prime examples, to strengthen the argument that it is reasonable to consider how far scientific detectives reflected or extended prevailing models of literary scientists. Second, the portrayal of scientific careers, or indeed lack of careers, in fiction revealed something of the trials and tribulations of would-be scientists in attaining professional status. Such difficulties apply to a wide range of scientific callings, of course; nevertheless they certainly applied to forensic science. The uphill battle to attain professional status and concomitant lack of professional opportunity were significant issues right up to the end of the nineteenth and well into the twentieth century right across the scientific spectrum.[12]

There is a moral side to the portrayal of fictional scientists; a consideration of this dimension weaves together scientific stereotypes and professional opportunity. Real life, *fin de siècle* and early twentieth-century men of science faced moral issues in terms of how the results of their work could be used to help others or not. Forensic scientists and medical men liked to portray themselves as sitting on the moral high ground, caring for truth rather than just the business of winning a court case. The lives of contemporary fictional scientists give considerable insight into the moral dilemmas that scientists faced, casting light on what Shapin describes as the 'moral equivalence' of scientists.[13] The term 'moral equivalence' refers to the dawning realization, probably stemming from late Victorian times, possibly earlier, that rather than portraying themselves as virtuous seekers after truth, caring little for worldly matters, scientific men were acknowledging that, in moral terms at least, they were just like everybody else. So the stories of men trying to carve out professional scientific careers were often far from heroic and noble. Couple this with fictional stories of 'mad and bad' scientists, and it is small wonder that many literary scientific stereotypes of this period were not always especially positive.

Scientists in fiction – making a living

The work of Haynes and Russell is especially relevant to a consideration of fictional scientific lives.[14] Haynes's focus is on scientific literary stereotypes, many of which were negative, and their ability to control themselves and their scientific creations.[15] Rather than considering the moral qualities of scientists *per se*, Russell is concerned with how far literature accurately portrays scientific careers and the difficulties of achieving a scientific living, particularly in the earlier of his two articles, which centres on the Victorian period.[16] Like Haynes, some of his message at least is pessimistic; it was typically very difficult for a lower middle-class man without substantial means to gain a suitable scientific education and a professional scientific appointment, in literature as in life.

Russell considers whether fictional scientists of the late Victorian era were able to make a living from a scientific career. As he notes, paid scientific appointments were still not plentiful in that period and it was hard for aspiring working-class or lower middle-class men of science to gain a foothold in the nascent scientific profession. The middle-class authors who included scientific characters in their novels, such as Hardy, Gissing and Wells, were themselves painfully aware of the limitations of class on professional aspirations.

A career in a routine role in an industrial laboratory does not sound like promising literary material, so it is no surprise that there are relatively few fictional examples. Nevertheless George Gissing's *Born in Exile* provides just such a tale, and not a happy one at that.[17] It is a story of a scientific career gone horribly wrong, and the 'hero', Godwin Peak, is a snob and a hypocrite. His lack of superior moral qualities leads to a career doing donkey work in an industrial laboratory and an early demise. It is a miserable tale, not just of the failure to obtain a glittering scientific career, but also of snobbery, hypocrisy and envy. Even if one feels sympathy for him, it is very difficult to like Godwin Peak. He is no scientific hero.

However much prominent Victorian scientists such as T. H. Huxley and John Tyndall argued for a secular scientific meritocracy, driven by ability rather than birth, the entry to a successful scientific career at the turn of the nineteenth and twentieth centuries was considerably eased by birth and social advantage.[18] Class was a more significant gatekeeper than talent for entry to a scientific career. Fictional scientific detectives were discreetly middle class and generally at least reasonably comfortable. At the beginning of *A Study in Scarlet* we know Holmes needed to find a partner to share the cost of his lodgings. As the canon progressed, there was occasional mention of a fee but never a very clear idea of Holmes's means and how far he needed those fees to maintain his modest middle-class life, not to mention his cocaine habit. John Thorndyke was a lecturer and barrister; he was qualified in law and medicine and was a medico-legal expert. He was clearly fairly comfortably off, and his fortunes contrast with those of his medically qualified partner, Jervis, his 'Watson' equivalent side-kick. Thorndyke could afford to employ a technician-cum-factotum in the form of Polton.

Even if professors of forensic medicine in ancient Scottish universities could have expected to make a very good living to the extent that they were not to be lured south for directorships of forensic science laboratories, we know that a medical training was far from being a passport to a comfortable living in the latter part of the nineteenth century as the careers of Conan Doyle and Austin Freeman attest. It is sobering to think that had Conan Doyle and Freeman flourished as medical practitioners, it is unlikely that Sherlock Holmes and Dr Thorndyke would have been born! Given first-hand experience that medical training was no easy route to financial prosperity, it is unsurprising that Doyle and Freeman reproduced such difficulties as a secondary theme in the fictional careers of Holmes's Watson and Thorndyke's Jervis. Indeed, when Thorndyke and Jervis initially re-encountered each other in the first Thorndyke mystery, *The Red Thumb Mark* (1907), almost the first thing Jervis did was to contrast the fortunes of the two medically trained friends, with Thorndyke in his gown and wig, working and living in salubrious chambers, whereas Jervis struggled on as a *locum tenens*.[19] Having been invalided out of his post in the army, Watson was glad to save money by sharing lodgings with Holmes in the first Sherlock Holmes novel, *A Study in Scarlet*.[20] Right at the beginning of the Holmes canon, Watson was portrayed as a medical man in somewhat straitened circumstances. In art as in life, a medical training may have facilitated a lucrative career, but it certainly did not guarantee it.

Some of the real-life forensic scientists who were born in the later years of the nineteenth century and whose professional lives spanned the early to middle years of the twentieth century tended to be drawn from the lower middle classes, insofar as sparse biographical details reveal. Firth described obtaining a scholarship to Stockport Technical School, whence he entered a pupil teachers' centre and attended evening classes in Manchester College of Technology in order to pass the matriculation examinations for entry to Manchester University and gain a scholarship to study chemistry.[21] From then on, his career took off; nevertheless, his does not sound like a particularly well-to-do background. Julius Grant came from a poor family in London's East End and, like Firth, started his early scientific education through evening classes while working full time in a chemistry laboratory during the day.[22] In contrast, Charles Ainsworth Mitchell, who was from a medical family, obtained an Oxford education.[23] It is impossible to generalize the early experience of those who became forensic scientists from so little data.

Nevertheless, it is reasonable to conclude that some mid-twentieth-century forensic scientists had to work their way up the career ladder, reflecting some of the difficulties fictional scientists faced.

Literary scientific stereotypes

Haynes's focus is on stereotypes of scientists in literature, rather than on the realism or otherwise of their careers, illuminating largely ethical or moral aspects of scientific behaviour instead. If there is a common theme uniting her stereotypes, then it is the question of control or lack of control. She distills the portrayal of scientists in literature into a set of seven primary stereotypes, including 'evil alchemist', 'noble, heroic scientist', 'foolish scientist', 'inhuman researcher', 'scientist as adventurer', 'mad, bad, dangerous scientist' and the 'helpless scientist' who cannot control the outcome of his research.[24] Her mad or bad scientists were unable to control themselves, their creations, the natural world or, indeed, all three. Fictional rather than factual scientists have influenced the negative cultural stereotype of the scientist as mad or malevolent, which is at odds with the views of many scientists who regarded themselves as working for social good.[25] However, the stereotype of heroic, scientific adventurer was marked by his ability to exercise control over himself in order to perform heroic, even superhuman, feats over his scientific creations, and certainly over nature; this stereotype is relevant to the scientific detective. At a time when scientific knowledge was regarded as the epitome of knowledge available to control and shape the natural world, the scientist's potential to control the latter was of supreme importance. There is little evidence of moral equivalence here as the scientific hero displayed virtue and courage well beyond the ordinary. The heroic adventurer scientist was not only achieving mastery over a sometimes hostile natural world; he (and it was 'he') was also achieving mastery over himself. Hence, his adventures and his morals are above and beyond ordinary morality.

Of course, control over nature and, more specifically, crime and criminals, by means of scientific knowledge was an essential part of the fictional scientific detective's repertoire, and this mirrored non-fictional concerns over control of crime. The scientific detective was instrumental in maintaining social order against perceived threats of crime and criminals represented as the 'other'. John Thorndyke was of this ilk. Sherlock Holmes was a slightly more ambivalent character. Of course, such true-blue portrayals of detectives would be revised by the later more 'hard-boiled' era of detection, with the ascendance of brilliant yet flawed detectives, relying more on human instinct and knowledge of human frailties and less on 'objective' science and cool rational thinking.[26] Late Victorian and Edwardian fictional scientific detectives generally displayed positive moral qualities. This works with Thorndyke, who was a fine upstanding moral crusader, but perhaps a little less well when one recalls Holmes's drug habit. There is a clear link between control and morality. The scientist who controlled himself and nature was displaying appropriate moral qualities to continue the scientific war of crime control.

Fictional scientists as adventurers

Jules Verne's Professor Lidenbrock in *A Journey to the Centre of the Earth* (1864) was the paradigm literary adventurer scientist.[27] Such scientists were bold heroic adventurers conquering the physical world, demonstrating progress and showing science as a force for good. This scientific type evolved into the hero of the optimistic science fiction story of the late twentieth century. Although not specifically addressing scientific detectives, Haynes classifies Sherlock Holmes in the 'scientist as adventurer' category. She characterizes Sherlock Holmes, alongside Conan Doyle's Professor Challenger, as exemplars of the type, 'with their potent mix of science, adventure, rationality, and moral superiority that restores justice and order out of evil and uncertainty'.[28]

This model is certainly more positive than Haynes's other scientific stereotypes which paint a picture of either a mad scientist or scientist not in control of himself or his creations. The maintenance of order is key. Importantly the heroic scientist was not just a brave adventurer; he was also morally superior. Indeed, scientific detectives generally fitted well into this mould with their superhuman powers of detection and their astonishing array of scientific competencies. Scientific detectives such as Dr John Thorndyke also looked the part. He is portrayed as tall, handsome and athletic as well as kind and genial.[29] As a follower of Herbert Spencer and a supporter of eugenics, Freeman naturally designed Thorndyke to be eugenically superior.[30] He appeared to have had no vices, unlike Sherlock Holmes. Holmes's cocaine habit would never have been entertained by clean-living Thorndyke. As Van Dover points out, Holmes's talent for disguise would also have been anathema to Thorndyke, as it was beneath the dignity of a gentleman to portray himself as something other than his true self.[31] Disguise was regarded as a form of deception and morally questionable, so detectives tended to 'play down its use in their rhetoric'.[32]

Somehow one cannot imagine Thorndyke playing games of 'witnessboxmanship' as in Walls's description of the art; possibly he did not need to do so as his talents were extraordinary.[33] He was supremely in control of himself and the case. Yet in Sherlock Holmes we do not see superhuman morality; he is morally superior up to a point. Despite the wide adulation his methods received, he was an ascetic, asocial figure who enjoyed the thrill of the detective case, the problem solving. Quotidian experiences bored him; without a case to solve he became depressed and turned to cocaine. One could regard Holmes's cocaine habit as displaying a lack of control over his own person, yet his drug use could be seen as deliberate rather than addictive. That he was morally somewhat ambiguous surely adds to Holmes's appeal. He worked tirelessly to fight crime, yet, for him, it was the intellectual exercise which was paramount and he makes no apology for it.

> 'My mind,' he said, 'rebels at stagnation. Give me problems, give me work, give me the most abstruse cryptogram or the most intricate analysis, and I am in my own proper atmosphere. I can dispense then with artificial stimulants. But I abhor the dull routine of existence. I crave for mental exaltation. That is why I have chosen my own particular profession,—or rather created it, for I am the only one in the world . . . The only unofficial consulting detective'.[34]

His admiration for the talents of his arch foe Moriarty was clear; he was a 'genius' with a 'brain of the first order' clearly a worthy match for Holmes's genius.[35] However, superior intellect and superior skills in mathematics and science did not necessarily mean that one would have superior moral qualities; indeed, when coupled with criminal tendencies, scientific skill amplified rather than tempered such tendencies, and here he alluded to contemporary Lombrosian ideas of hereditary criminality even if he did not describe them thus. Moriarty was a professor of mathematics gone bad, with 'hereditary tendencies of the most diabolical kind. A criminal strain ran in his blood, which, instead of being modified, was increased and rendered infinitely more dangerous by his extraordinary mental powers'.[36]

Scientific detectives, even if they could be seen as heroic, were generally not adventurers in the Jules Verne or Professor Challenger sense. In solving crimes, often, but by no means always, murders in London or elsewhere in England, they preserved a level of realism even when plot and scientific techniques employed for detection stretched that realism considerably. In the halcyon days of widespread, apparently effortless, train travel on minor railway lines in England, Holmes was more likely to travel to Boscombe Valley by train than round the world in eighty days.[37]

Sherlock Holmes, scientist or not?

Although Edgar Allan Poe's 'The murders in the Rue Morgue' and Wilkie Collins's *The Moonstone* are generally taken to be the first detective stories to be widely read, Conan Doyle's Sherlock Holmes was the first literary scientific detective of note, and he is the one who set the standard for all literary scientific detectives who follow.[38] That there are several works which examine Holmes's career in relation to science, which are discussed later in the chapter, is not especially surprising given the vast array of 'Sherlockiania' which has been spawned. This includes the *Baker Street Journal*, a group of 'Baker Street Irregulars' named after the street urchins who occasionally helped the master, museums, artefacts, parodies and many other literary productions too numerous to mention.[39] Sherlock Holmes remains a huge cultural industry, with his scientific accomplishments a significant contribution to that industry. Given his considerable influence on cultural life in general, and on forensic scientists in particular, his application of science to the solutions of crimes demands consideration.

Given the enthusiasm for Holmes's character and his scientific accomplishments, how far did the Holmes canon demonstrate the potential of new scientific techniques in crime detection? Certainly, the enthusiasm of several authors suggests that it does. However, the influence is better read in terms of something more subtle. Coupled with the interest in Holmes as a 'proto' forensic scientist, it is the possibilities of a scientific method of detection that Holmes seemed to offer rather than tangible techniques, which explains his scientific appeal, and this is borne out by Steenberg's view that Holmes acted as a kind of origin myth for forensic science. Yes, there were allusions to chemical experiments – Holmes was in the laboratory discovering a novel technique for identifying blood when he first met

Watson in *A Study in Scarlet*. However, closer inspection reveals that Holmes, it seems, was not quite such an accomplished scientist as later critics would like him to have been. At least, he never attained the convincing display of experimental scientific prowess which Thorndyke demonstrated, and he solved crimes by scientific *deduction* rather than scientific *analyses*. Even his attention to the crime scene – note his reference to the 'herd of buffalo' trampling over the scene in the Boscombe Valley Mystery – was hardly at the level of detail of Gross's or Lucas's descriptions of crime scene management.[40]

When Doyle created Sherlock Holmes, he wanted to 'show that the general lines of reasoning advocated by Holmes have a practical application to life'.[41] Not surprisingly, he exploited his own scientific interests developed during his medical training and career. His 'pilots' were T.H. Huxley, John Stuart Mill and Herbert Spencer.[42] He drew on the considerable cultural capital which science had achieved by the end of the nineteenth century, not least in terms of its application to criminal justice in the form of scientific criminology. Holmes's twin powers of deduction, based on acute observation, and his related scientific skills were woven together to produce the right kind of detective story. Although his powers of deduction continued undiminished throughout the Holmes canon, scientific analyses appear predominantly in the first half, before his famous (near) fatal encounter with Moriarty. As O'Brien notes, Holmes seemed to lose something of his scientific 'mojo' after the fall over the Reichenbach Falls. Doyle reluctantly brought back his most famous creation after this incident in response to immense pressure from Holmes's fans, but Holmes was never really quite the same, at least in terms of scientific prowess.[43]

A number of critics have argued that the apparent realism of Holmes's powers of deduction has been significantly exaggerated.[44] Holmes was not the superhumanly logical machine that Conan Doyle would have us believe. Instead, his deductions involved on many occasions extraordinary mental leaps. Indeed, Holmes, in *The Hound of the Baskervilles*, referred to 'the scientific use of the imagination', alluding to John Tyndall's famous lecture of the same title.[45] Anyone who claims in real life to follow Holmes's style of reasoning would be producing deductions which would hardly stand up in a courtroom. I have suggested that an allusion to Holmes was popular amongst those practising the forensic arts. However, more astute critics, Sydney Smith, J. B Firth, Keith Simpson and Isaac Asimov included, understood the scientific limitations of the Holmes canon.[46]

However, Conan Doyle's exceptional talent as a writer was to make Holmes's extraordinary powers of reasoning appear just about believable. And it is our willingness to suspend disbelief about Holmes's powers of deduction that fuels some of the more enthusiastic claims of his contribution to the founding of forensic science from practitioners such as Berg.[47] Maintaining that Holmes invented the subject is a rhetorical strategy. Basing one's claims to professional legitimacy on a fictional character needed careful handling. At first sight this could detract from rather than enhance the professional status of those in the nascent forensic science profession. Would one model oneself on a distinctively dressed fictional late Victorian eccentric? However, it is the very success of Holmes in solving crimes with 'scientific deduction', coupled with the fact that he was such a famous cultural referent, which meant that drawing on his approach to scientific crime solution offered considerable cultural purchase and a promise of legitimation.

Dr John Evelyn Thorndyke

Van Dover argues that Freeman's special skill lay in combining 'a sober method and a clever but uninebriated narration'.[48] Freeman explained Thorndyke's method rather than relying on a 'presumption of method' as others did. Whereas we know little of Holmes's qualifications, Thorndyke held an M.D. and D.Sc. and this makes his profession different from Holmes, who was a freelance consulting detective; Thorndyke was a medico-legal expert or medical jurist, a trained doctor and lawyer.[49] Thorndyke told Jervis at their first meeting he

> 'appeared in the character of that *bête noir* of judges and counsel – the scientific witness. But in most cases I do not appear at all; I merely direct investigations arrange and analyse the results, and prime the counsel with facts and suggestions for cross-examination'.[50]

Holmes operated just outside the establishment, but Thorndyke, as a barrister and expert witness, was very much inside it.[51] He did not have Holmes's bohemian inclinations, but he did have scientific imagination in great abundance.[52] His methods were different from those of Holmes in that they were more technical, more focused on the 'interrogation of things rather than persons'.[53] That Freeman verified every experiment he attributed to Thorndyke serves to add considerable scientific authority to the character; however, this did mean that some stories relied on specialist knowledge that a reader could not possibly know.[54]

Although Thorndyke is not nearly as well known as Holmes, he has quietly retained a popular base in the UK over the years with a television series in 1964 and recent radio appearances. One of these radio adaptations dubs him 'forensic investigator'. The term medico-legal investigator is preferred in the Thorndyke canon, so it seems as if Thorndyke has been reimagined as a forensic scientist for the twenty-first century.[55] Thorndyke only occasionally used the adjective 'forensic', referring to something connected with the law by this use of the term. The term, 'forensic medicine' is occasionally mentioned but 'forensic science' is not a term he would have used. Nevertheless, his grasp of laboratory sciences means that he can easily be enrolled as a kind of proto–forensic scientist.

Thorndyke was the creation of Richard Austin Freeman, a near contemporary of Conan Doyle. Thorndyke's adventures are set in London and contain much more scientific detail, indeed convincing scientific detail, than the Holmes stories; the plot often hinges on a scientific denouement. Freeman stated that he consciously wished to include realistic scientific analysis even though something of the pace of the plot might be sacrificed to the science.[56] One of the techniques used by Freeman was the inverted story. In this, the reader witnesses all the stages of the crime and then is led back through the stages of Thorndyke's detective process to the apprehension of the criminal. One might say the interest lies not in *whodunit* but in *howsolvedit*. The inverted story never became a very popular style of crime writing, possibly because the suspense is somewhat diluted by the story's structure, but it did show how much of an emphasis Freeman was willing to place on scientific detection to provide the main narrative interest.

It is salient that Thorndyke arrived on the scene some twenty years after Holmes and the last Thorndyke work appeared in 1942.[57] This makes Thorndyke a creature of a different era; he lived at a time when forensic techniques were rapidly being developed and where the new state laboratories were created. Unlike Holmes, he appeared in court and was well versed in the art of holding one's own in cross-examination. For instance, in *The Mystery of Angelina Frood* he knew that reliance on Lucas's *Forensic Chemistry* would be acceptable in a coroner's court.[58] In *The Shadow of the Wolf*, he examined a section of a mineral which was crucial to the case under a microscope (the section, of course, was provided by Polton). Although he was able to work out the composition of the mineral, he also knew to have this verified by a friend, an expert from the Geological Museum, as he knew the latter's specialism meant the evidence would stand up in court.[59] As Donaldson noted: 'As a sound jurist, Thorndyke is aware of the value of corroborative evidence from an expert, even in those cases where he is confident of his own knowledge.'[60] Indeed, in *A Silent Witness*, he assembled a team to help him in 'the most convincing laboratory-workshop scene ever described'.[61]

The best way of understanding the influences surrounding the literary birth of John Thorndyke is to read Freeman's own description of his influences in 'Meet Dr Thorndyke'.[62] As a medical student, Freeman was impressed by the dramatic aspects of medico-legal cases. He was much influenced by Sherlock Holmes but wanted to create a detective story 'based on the science of Medical Jurisprudence, in which, by the sacrifice of a certain amount of dramatic effect, one could keep entirely within the facts of real life, with nothing fictitious excepting the persons and events'.[63] Unlike Conan Doyle's modelling Holmes on his old lecturer Joseph Bell, Freeman did not have a real person as a model, although Donaldson suggests that Thorndyke owed much to the character of Alfred Swaine Taylor, whom Freeman would have known only through Taylor's textbook.[64] After all, Freeman acknowledged the influence of his textbooks, especially the cases used for illustration: 'I was profoundly impressed by their dramatic quality'.[65]

Polton, however, had links to real people – the laboratory assistant in the hospital museum where Freeman was a student and a watch maker 'who was a man of boundless ingenuity and resource'.[66] Van Dover notes that the distinction between Polton and Thorndyke was one of 'craftsman and scientist'; a matter of class.[67] Polton was an ingenious maker of things and this fact considerably aided Thorndyke so that he was able to incorporate a use of technology in ways that Holmes could not.[68]

Holmes, Thorndyke and fingerprints

Leaving aside Holmes's powers of scientific deduction, how did he apply science to crime; in other words, in what ways did he act as a forensic scientist, and how did this compare to Thorndyke's scientific skill? Space permits comparison of only one aspect of forensic skill; fingerprint analysis, the gold-standard forensic identification technology, provides an important example. Holmes admired Bertillon's system: 'His conversation, I remember, was about the Bertillon system of measurements, and he expressed his enthusiastic admiration of the French savant.'[69] Nevertheless, he was offended by James Mortimer's ranking of Bertillon as the preeminent European expert in *The Hound of the Baskervilles*.[70] Despite his avowed admiration and despite the fact that *Bertillonage* was very much *au courant* when Holmes started out, Holmes never appeared to make use of it.

Fingerprinting did appear in the Holmes canon. O'Brien cites fingerprinting appearing in seven of the sixty Holmes stories.[71] Given that fingerprinting would rapidly become the *de facto* forensic identification technique after Scotland Yard's adoption of the method instead of *Bertillonage* in the early twentieth century and there was considerable discussion in the pages of *Nature*, through Galton's book on fingerprinting and elsewhere, it is perhaps surprising that Conan Doyle did not make more use of the technique in the Holmes canon.[72] Fingerprint analysis would have had a considerable level of novelty at the time and, given its rapid ascendance in the armoury of forensic technoscience, it would have represented supreme scientific currency.

Even before fingerprinting was adopted officially in criminal detection and came to be seen as an infallible forensic technique – in other words, even before there was a working system – it was widely believed that fingerprints were likely to be unique and so could be used for criminal identification and record keeping. Mark Twain was an 'early adopter' of fingerprinting as a literary device. The denouement of his book *Pudd'nhead Wilson* (1894) rested on a fingerprint left at the scene of a murder.[73] Under these circumstances it is interesting that fingerprints do not appear more often in the Holmes stories. Indeed, O'Brien points out that in several stories fingerprints were noticed by Holmes but then they were not used to catch a criminal.[74]

In 'The adventure of the Norwood builder' the case hinged on a fingerprint; indeed, a forged fingerprint which the perpetrator of the crime obtained from an impression on a wax seal.[75] This may well be the first time a false fingerprint is used in literature, although it was certainly not the last, as the plot of Austin Freeman's first Thorndyke mystery, *The Red Thumb Mark*, rests entirely on the purported ease with which fingerprints may be forged.[76] Comparing the use of forged fingerprints as a plot device in the two stories reveals that they are completely central to the Thorndyke story. The forged fingerprint is more peripheral to the plotline in 'The adventure of the Norwood builder'; there is little discussion of how the forgery was made and there is no evidence that Conan Doyle either experimented with copying fingerprints or concerned himself with the technicalities.[77] Holmes regarded it as a simple matter to make the forgery from a thumb impression in a wax seal:

> It was the simplest thing in the world for him to take a wax impression from the seal, to moisten it in as much blood as he could get from a pin-prick, and to put the mark upon the wall during the night, either with his own hand or with that of his housekeeper.[78]

However, in Freeman's case, if he used the plotline of a forged fingerprint, it was because he had tried out the techniques for producing such a print and had confirmed that these techniques worked sufficiently well to translate into fiction.

Indeed, in his preface to *The Red Thumb Mark*, Freeman explicitly stated that the aim of his book was to educate, namely to show how easy it is to forge a fingerprint.[79] However, if anything, Freeman demonstrated that it is actually quite difficult to forge a fingerprint, although it is eminently possible, as it requires considerable knowledge of the chemical and physical properties of blood. Despite the significant technical skill involved in creating a convincing forged fingerprint, Freeman wanted lay and professional audiences to understand that fingerprints *could* be forged and that freely making one's fingerprints available in the fingerprint equivalent of autograph books, something that was popular early in the twentieth century in the form of 'Thumbographs', made one's fingerprints available to a potential forger.[80]

The Sherlock Holmes and science meta-canon

Rather than considering the content of the Holmes canon for uses of science, it is much more appropriate for the purposes of this work to consider what might be termed the Holmes scientific meta-canon – in other words, the writing on his scientific credentials which the Sherlock Holmes phenomenon has inspired. The contemporary works in the meta-canon of Holmes literature, which are based entirely on the connection between Sherlock Holmes and science, bear witness to the oft-quoted link between Holmes's detective methods and scientific analyses. Four such recent works are briefly reviewed here – all have been published in the last fifteen or so years. When I initially tracked down these books, I was surprised by their contents. It was years since I had originally followed the adventures of Sherlock Holmes and, like everyone else it seems, I remembered them as revealing the arts of scientific detection and the inspiration of later forensic techniques. I anticipated that these works would tell me exactly where the science was in Sherlock Holmes and would be definitive in cementing the exact ways in which the Holmes canon relies on scientific analyses in fighting crime. However, despite the fact that these are all well-researched works and I have gone over them with my literary magnifying glass, I believe that none succeeds in making a convincing link. These books achieve a somewhat different objective. Rather than making a convincing connection between Holmesian detection and exact scientific analyses, their potentially weaker claims of references to science in the Holmes novels are coupled with discussions of real-life contemporary crimes, criminal detection and science in general. What they achieve, rather, is a positioning of the Holmes literary edifice against a sympathetic, non-fictional scientific and forensic landscape. This reinforces the contention that it is the common cultural imaginary surrounding science and the detection of crime which is important in setting the scene for the emergence of forensic science as a set of scientific disciplines in the twentieth century – something to which Holmes and the other literary scientific detectives contributed and something which, it seems, is considerably more important than nailing down a demonstration of Holmesian scientific expertise.

Holmes, Chemistry and the Royal Institution is based on a lecture given to the Sherlock Holmes Society of London at the Royal Institution (RI) in 1994.[81] The book contains much peripheral material on the Royal Institution, a list of RI discourses which Holmes might have enjoyed if he had been a real person attending the lectures, a comparison of Holmes's and Faraday's characters, discussion on where Holmes learned his chemistry and the history of science in general. His monographs, study of cigar ash and oft-quoted 'profound' knowledge of chemistry are all cited. The chapter 'Sherlock Holmes: A study in chemistry' details the famous first meeting of Holmes and Watson where Holmes has just discovered his test for haemoglobin, a test which would potentially have revolutionized the detection of some crimes but which did not reappear in any of the subsequent Sherlock Holmes stories.[82] Richards and Gore note an example where Watson remarked that Holmes has been at work again with his tubes and chemicals.

> Watson asks if Holmes had solved the case of Miss Sutherland, to which Holmes, thinking that Watson is referring to his chemical work, replies that it was bisulphate of baryta. As to the significance of the work, the reader is not informed, but it can be inferred that this result helped bring another criminal to justice.[83]

Conan Doyle provided no explanation as to the relationship between Holmes's experiment and the solution to a crime; indeed, the fictional experiment may have had nothing whatsoever to do with solving a crime. However, Richards and Gore express the commonly held view that it must have borne some relationship to solving crime, and it is this belief which appears to be more important than the link itself.[84]

Lindsey's *Sherlock Holmes and a Question of Science* (2006) attempts to illustrate the wide-ranging scientific and medical references in the Holmes canon.[85] This book has a more 'encyclopaedic' cast than the other Holmes and science works reviewed here, as it lists references to medical practitioners, poisons, firearms, etc., in short paragraphs without attempting to sustain a critical narrative. As Lindsey notes, Holmes used his chemical experiments as a kind of therapy. Tellingly he argues:

> Holmes' use of analytical chemistry certainly helps him to solve crimes, and at the same time increase his knowledge of this particular branch of science which so fascinates him. In this way he carries out valuable research, as exemplified by his experimental test for blood in **A Study in Scarlet**. By doing so, he hopes to push forward the boundaries of this new (in the 19th century) discipline, which today is known as 'forensic science.'[86]

Indeed 'modern laboratory methods may well have obtained a result in **The Speckled Band**, when at the end of the investigation, Holmes comments on the cleverness of the murderer in "*using a form of poison which could not possibly be discovered by any chemical test*"'.[87] Lindsey argues that Holmes's chemistry definitely helped him to solve crimes, although there are very few clues to how this was achieved in his book and in the Holmes canon, and he imputes in Holmes the express wish to push forward forensic science without evidence for the claim. This is another example of our willingness to let Holmes off the hook of having to definitively explain and connect his science to solving a crime.

The title of Wagner's book, namely, *The Science of Sherlock Holmes: From Baskerville Hall to the Valley of Fear, the Real Forensics Behind the Great Detective's Greatest Cases*, promises to reveal a clear link between Holmes's scientific analyses and his solving of crimes.[88] The book opens with a brief review of the history of forensic medicine. The second chapter, 'Beastly tales and black dogs', reflects on myths of black dogs in British culture. Subsequent chapters cover a wide range of forensic sciences and technologies, well-known US and European forensic scientists and major cases such as Crippen and the Tichborne Claimant. This is a very comprehensive coverage of the forensic milieu of the broad period – it is a history of forensic science – but the links with the science of Holmes are tenuous.

The title of O'Brien's more recent work, *The Scientific Sherlock Holmes: Cracking the Case with Science & Forensics*, appears to make a similar promise.[89] The first two chapters review the main characters in the Sherlock Holmes canon. This sets the scene rather than making scientific connections. The book's third chapter is titled 'Sherlock Holmes: Pioneer in forensic science'. O'Brien notes that Holmes was an admirer of the Bertillon system but never used it in his adventures; similarly, fingerprints appeared but, except for 'The Norwood builder' they were not used catch a criminal.[90]

In fact, Holmes was far more interested in footprints than fingerprints. According to O'Brien twenty-six of sixty cases mention these, and Holmes appears to have had some success in identifying suspects from footprints.[91] Interestingly, O'Brien dismisses footprint analysis as of 'limited utility' in modern crime detection because of mass production of shoes. However, this ignores minor but detectable variations in shoe sole patterns and the unique pattern of shoe sole wear, which everyone produces, the huge variation in shoe soles and the fact that we all have to place our feet on the floor or ground, all of which serve to make analysis of shoe marks of forensic relevance. Contra O'Brien, Holmes *was* anticipating later forensic science in his interest in shoe prints. He used a microscope in his chemical research, although it is his famous magnifying lens for which he is best known. The one clear example of Holmes definitively using a microscope for forensic purposes, also referenced by Lindsey, occurred in 'The adventure of Shoscombe Old Place' (1927) where he found fibres from a tweed coat, dust, glue particles and skin cells and where he remarked that the police had begun to be interested in the use of microscopes.[92] O'Brien's chapter on Holmes as chemist is interesting as it focuses on the period when Holmes disappeared after the Reichenbach Falls incident. Although he is at pains to refute in some detail the view of Isaac Asimov that Holmes was a blundering chemist, O'Brien notes: 'Rather than being the focus of a case, his forensic chemistry was sometimes merely mentioned by his chronicler Dr. Watson.'[93]

These four works promise in varying degrees to uncover Sherlock Holmes's relationship to forensic science; however, they all achieve something different, namely a catalogue of relevant science in the real world and related real-world cases and science. Only O'Brien's book gets close to uncovering Holmes's contribution to the development of forensic science, but this is not in relation to chemistry. For all Holmes's interest in the subject, his all-night experiments and his discovery of a haemoglobin test that was never used, he does not use chemical analyses in the detection of crime. He does not use the contemporary technique *Bertillonage*, nor does he make much use of fingerprinting which had become the gold standard by the end of his career. He does appear to have used footprints quite effectively, and he understood the importance of preserving a crime scene. Although scientific analyses and the writing of scientific monographs are alluded to in the Holmes stories, they appear as epiphenomena. If we were to base the appeal of the Holmes stories on their role as harbingers of forensic science techniques, it would be difficult to understand their fascination.

In this respect, Holmes was unlike Thorndyke, who did make use of scientific techniques even if narrative action is, to some extent, sacrificed. Importantly, Thorndyke employed an expert technician in the form of the faithful Polton. Polton is much more important than we might first imagine. Van Dover regards Thorndyke as a model of gentlemanly respectability and propriety; Thorndyke, the gentleman, and Polton, the servant, are clearly delineated by class. However, the creation of Polton was a very significant way in which Austin Freeman acknowledged the knowledge and expertise involved in creating the scientific analyses which Thorndyke used in his stories. Thorndyke may have been the only fictional scientific detective, or more accurately, fictional medico-legal expert, of that era who employed a technician in this way, and his stories are all the more scientifically convincing for it. It is Polton who provides much of the scientific wraparound – in other words the technoscience – which makes Thorndyke's forensic career convincing. Freeman tried out the forensic techniques which appeared in the Thorndyke stories, even to the extent of using some of his own photographs of trace 'evidence', so they are much more detailed, convincing and central to the plot; he knew the difficulties involved.[94] This appearance of realism is at least part of the reason why Thorndyke is the most convincing scientist amongst literary scientific detectives. The difference in the level of description of the means of forging fingerprints in *The Red Thumb Mark*, several pages describing Thorndyke's evidence in court, compared with the brief description of the print being forged in 'The Norwood builder' attests to this.[95]

Although I have alluded to the nods that forensic scientists made towards Holmes, some were openly critical of his approach. As Firth argued, Holmes never had to appear in the law courts; in this respect he is different from Thorndyke who often appeared in court as an expert witness to explain his analyses.[96] Forensic pathologist, Keith Simpson pointed out:

> Conan Doyle never allowed Sherlock Holmes's 'profound' knowledge of chemistry to reveal itself in practice. It would never have made good reading

> anyway: the brilliant storyteller would have lost his way in the scientific mazes created by the atomic scientist and x-ray crystallography.[97]

His explanation for the lack of real science in the Holmes stories was that 'to rationalize and detail might well have destroyed the magic simplicity of Sherlock Holmes's skills in deduction'.[98] Other forensic scientists pointed to Thorndyke as a more convincing disciple of forensic science. Charles Ainsworth Mitchell believed that Sherlock Holmes's deductions were 'supposed to be based on accurate observations' but they were sketchy and lacking scientific probability.[99] By way of contrast, Mitchell saw Thorndyke as far more convincing, as the facts and scientific deductions were described step by step and only rarely, and in a minor way, were there any errors in Thorndyke's science. Mitchell found Thorndyke to be plausible and good for demonstrations of a number of scientific techniques associated with solving crime. As already noted, Lucas was friendly with a number of crime writers and inspired some crime fiction writing.[100] Thorndyke appears to have been the forensic scientist's fictional forensic scientist.

However, the point at issue here is not to claim that one fictional scientific detective is a better scientist than another. Sherlock Holmes eclipsed all others in literary power, including Thorndyke, even if he was less of a scientist than Thorndyke, and it was the 'scientific method' which Holmes uses which contributes so effectively to his popularity. Even if his deductions, when inspected more closely, involve unwarranted speculations and inferences, the miasma of scientific method helped him achieve popularity. Rather than asking, 'Where's the science?' we should look to Holmes as reflecting the wider cultural imaginary which linked forensic matters with science. A link can be convincingly forged between Holmes and the 'scientific aids' movement and scientific detection of the 1920s and 1930s in the UK in terms of the promise of a scientific method and the rhetoric of 'scientific' as modern, progressive and controlling, a point which the Sherlock Holmes adventures *and* the more quotidian 'scientific aids' publications all make. They are both essential elements in the performance of 'scientific' in forensic matters.

Steenberg, in her analysis of contemporary forensic culture, points to the importance of mythical, historic beginnings to forensic sciences. In particular, she notes:

> Popular stories about forensics generally frame their histories using an evolutionary model, seeing the (British) nineteenth century as the crucible in which its genre, technologies and disciplines were forged. The late nineteenth century of Sherlock Holmes and Jack the Ripper serve as kind of nostalgic 'golden age' in the metanarrative of forensics . . . in return to the Holmes and Ripper stories, the forensic sub-genre is talking to itself about itself and forensic discourse is cementing its own mythologies.[101]

The so-called *CSI* effect, or the idea that fictional crime dramas such as *CSI* affect our understanding and expectations of forensic evidence and those who deal with such evidence, could just as well be labelled the 'Sherlock Holmes effect', given the allusions of many textbooks of forensic science to Sherlock Holmes in relation to the development of forensic science and technologies.[102]

So Sherlock Holmes is burdened, perhaps over-burdened, with the weight of being the fictional progenitor of forensic science even though he offered little by way of a realistic role model. He was middle class but we hear little about his background, education and financial means. There was no professional or institutional wrap-around. Although he was concerned with the preservation of the crime scene, he does not have the obsession with the minutiae of packing which so concerned the forensic scientists who were working towards the end of his career. He did not require a lab technician, or for the most part, a specialist laboratory, no doubt driving Mrs Hudson to distraction with his table full of chemicals and glassware. He was a freelancer with a sometimes difficult relationship to the police, and he never appeared as a witness in court, one of the activities which scientific expert witnesses of the early British forensic laboratories found so testing, and which was a skill which had to be learned as some of the forensic autobiographies revealed. Thorndyke, as a university lecturer and trained barrister, who presented forensic evidence in court and had the valuable services of a much-needed technician, offered a more substantial role model. So the important differences between Thorndyke and Holmes lie less in who was the better scientist. In demonstrating a nascent professional laboratory setting with trained technical staff and by showing what it was like to present detailed scientific evidence in court, i.e., acting as scientific expert witness, Thorndyke was a more convincing role model for the early twentieth-century forensic scientist.

Life imitating art imitating life

Holmes's influence on later forensic scientists is legendary. It is striking how Holmes becomes not almost real, but real, in this mythology. The UK's National Archives hold a thick file of letters sent to Scotland Yard enquiring as to the whereabouts of Holmes, and Conan Doyle was frequently asked whether he was real.[103] Indeed, Conan Doyle recounted in his memoirs:

> I do not think that I ever realized what a living actual personality Holmes had become to the more guileless readers, until I heard of the very pleasing story of the char-à-banc of French schoolboys who, when asked what they wanted to see first in London, replied unanimously that they wanted to see Mr. Holmes' lodgings in Baker Street.[104]

221B Baker Street, London, with its blue plaque indicating when Holmes lived there, now houses a Sherlock Holmes museum preserving the famous study, which, of course, never existed![105]

If readers were determined to make Sherlock Holmes real, they also tried to turn Conan Doyle into Holmes. For instance, Conan Doyle became involved in attempts to clear the name of George Edalji, a young solicitor who was charged with maiming animals in the Great Wyrley area of Staffordshire in 1903. As a trained opthalmist, his first meeting with George convinced him his eyesight was so poor he could not possibly have committed the crimes of which he was accused.[106] He visited the crime scenes, interviewed local residents and pointed out a number of errors in the analysis of crime scene materials, errors which Holmes would clearly not have made. For instance, the mud on Edalji's boots was of a different type from that of the fields where the maimings took place; indeed, the maimings continued after George was imprisoned.[107] In response to Conan Doyle's research, the *Daily Telegraph* badged its editorial on the topic: 'Sherlock Holmes at work'.[108] Eventually, Edalji was found innocent of the maimings. Conan Doyle's influence was clearly considerable in establishing his innocence.

No one seems to have been more convinced as to Sherlock Holmes's credentials for being the inventor of a spectrum of forensic techniques than mid-twentieth-century ballistics expert Stanton Berg, contending that 'a strong case can be made that the famous sleuth had a decidedly stimulating influence on the development of modern scientific crime detection'.[109] He named a raft of forensic scientists of the middle years of the twentieth century who claimed Holmes as their inspiration, to the extent of claiming that some forensic techniques which were to become common in crime laboratories in the middle years of the twentieth century were invented by the famous sleuth – rather taking at face value Holmes as an origin myth. According to Berg, Bertillon believed that Holmes's methods should have been adopted by all police forces. Locard is supposed to have first realized the forensic importance of dust through the influence of Holmes.[110] Berg suggested that Doyle may have got his ideas from Gross, or perhaps it was the other way round, although there is no way of knowing, given that Sherlock Holmes first appeared in 1887 long before the first English translation of Gross's *Handbuch.* A number of forensic scientists enjoyed the accolade of being called Sherlock Holmes. For instance, Edmond Locard was apparently known as Lyon's 'Sherlock Holmes', and Alfred Lucas was 'Egypt's Sherlock Holmes'.[111]

Thorndyke did not enjoy quite so much acknowledged real-life influence as did Sherlock Holmes; nevertheless, just as with Sherlock Holmes and his author, Conan Doyle, we are occasionally enjoined to imagine influences on real forensic scientists and medico-legal experts. The crime writer Raymond Chandler was an enthusiastic fan of Thorndyke. He credited Freeman with 'proving the possibility of forging fingerprints and of detecting the forgeries before the police thought of such a thing.'[112] It is interesting that Chandler either forgot that 'The adventure of the Norwood builder' used forged fingerprints first, or perhaps he did not regard the Holmes story, with its minimal allusion to the method of forgery, as providing proof that fingerprints could indeed be forged. He invited us to imagine Spilsbury and Thorndyke meeting in court. 'The great scene would have been a court-room battle between Thorndyke and Spilsbury, and for my money Thorndyke would have won hands down.'[113]

Real-life and fiction became mixed in interesting ways in the forensic arena. Forensic chemist Alfred Lucas corresponded with various writers of detective fiction and was a particular fan of Austin Freeman's Thorndyke novels.[114] It is notable that Freeman used Lucas's demonstration of the preservative powers of quicklime on a body in his book *The Mystery of Angelina Frood.*[115] This discovery had been an important part of the analysis of the Crippen case, explained rather after the event, as it helped to clarify the body tissues' good state of preservation.[116] The denouement of *Angelina Frood* hinged on the fact that skeletal remains discovered in a wall, presumed to be those of the eponymous Angelina Frood, where the flesh was believed to have been destroyed by quicklime, could not have been the lady in question.[117] Thorndyke's evidence to the coroner's court hinged on Lucas's experiments with preserving bodies in quicklime. These showed that the assumption of Pepper, the pathologist, in the Crippen trial of 1910, namely that quicklime destroys human flesh, was wrong.[118] Thorndyke described Lucas's experiments and offered a copy of Lucas's *Forensic Chemistry* to the court as evidence. The admissibility of the evidence was questioned. 'The witness can't swear to another man's experiments.'[119] However, as it was a coroner's court, they were not bound as strictly to the rules of evidence as in a criminal court. In any case, Thorndyke had repeated Lucas's original experiment, using skinned rabbits rather than the plucked pigeons of the original experiment, and had confirmed his results, namely that quicklime produces a desiccating, mummifying effect that preserves the flesh and hinders putrefaction.[120] Of course, such was Freeman's extraordinary commitment to forensic veracity that he, *himself*, had undertaken the rabbit experiments before writing the book.[121] Interestingly, *The Mystery of Angelina Frood* was first published in 1924 and Wilcox's lecture on the Crippen case, where he acknowledged the preservative effects of quicklime, was delivered in 1924.[122]

Of course, forensic experts were from time to time called upon to advise crime fiction authors. John Glaister Jnr explained in his autobiography: 'I've had many an urgent appeal from a crime writer whose plot appeared in danger of becoming unstuck on a point of forensic probability.'[123] Indeed, Glaister corresponded with Erle Stanley Gardner, creator of the fictional defence lawyer, Perry Mason, advising him on forensic techniques and cases.

> Yet how many of the millions of Perry Mason fans would guess that the lawyer-detective's adventures against the romantic, sunlit background of California are often based on facts which emerged amid the smoke-grimed tenements of Glasgow or some rain-soaked Highland hill?[124]

Glaister's involvement with shaping fictional forensics did not end there, as he was an advisor to the popular BBC drama *The Expert*, based on the adventures of a Home Office pathologist. The series was produced by his nephew, Gerard Glaister, and was broadcast in sixty-two episodes by the BBC between 1968 and 1976.[125]

Fictional forensic equipment also appeared in the 1924 English translation of Gross's *Handbuch*, as Burney points out.[126] The editor, John Collyer Adam of Madras, inserted a reference to Thorndyke equipment, namely Thorndyke's portable forensic laboratory. As this was the second English edition, based on the 1903 German-language edition of the *Handbuch*, and Thorndyke did not appear until 1907, it is definitely Adam's addition. It referred to a version of the portable kit which Gross had advocated in his original work.[127] In the third English-language edition, published in 1934, the reference to Thorndyke equipment had gone.[128] The reference to Thorndyke equipment was not to reappear in later English-language editions of the *Handbuch*, possibly because fact was catching up with fiction. By this time, the 'scientific aids' movement was gathering pace and the police were devising their own versions of portable forensic laboratories. A prime example of a portable laboratory worthy of Thorndyke was described in an article appearing in the *Police Journal* in 1934. Detective Inspector Pentland of the Nottinghamshire Police published a description of his 'Detective's Portable Forensic and Finger-Print Outfit', which could be carried to the crime scene in a small case.[129]

Conclusion

The literature of scientific detection influenced the development of thinking about forensic science in popular culture. This influence must be set against a backdrop of the portrayal of scientists in late Victorian and Edwardian literature, which was a mixture of positive and negative, both in terms of scientific careers and in terms of the control that fictional scientists exerted over themselves and over the natural world. Scientific detectives provided potential inspiration for the role of the forensic scientist, with several contemporary authors attempting to forge a strong and definite link between Holmes and scientific analyses in the detection of crime. Despite the acknowledgment that many forensic scientists seemed obliged to make to Sherlock Holmes, astute forensic practitioners knew that Thorndyke was a more convincing fictional forensic scientist.

But this was all part of a common cultural imaginary surrounding the development of forensic science, against which fiction and real life were woven together. Detective fiction completes the picture of the British beginnings of forensic science which moves from the construction of scientific and legal epistemological categories and the introduction of scientific expert witnesses in court, to the influences of scientific criminology and criminalistics in the UK and in the empire, the establishment of forensic science laboratories and forensic careers and the self-image that forensic biographies portrayed. Finally, the reception of scientific detective fiction and its extraordinary appeal is woven into the story of the origins of forensic science in the way that authors of scientific detective fiction acted as forensic scientists and real-life forensic scientists made their appearances in fiction.

Notes

1 C. Lindsey, *Sherlock Holmes and a Question of Science*, Leatherhead: Hadley Pager Info, 2006; L. Panek, *The Origins of the American Detective Story*, Jefferson, NC: MacFarland & Co., 2006; A. J. Richards and B. Gore, *Holmes, Chemistry and the Royal Institution*, London: Sherlock Holmes Society of London, 1998; R. R. Thomas, *Detective Fiction and the Rise of Forensic Science*, Cambridge and New York: Cambridge University Press, 1999; E. J. Wagner, *The Science of Sherlock Holmes From Baskerville Hall to the Valley of Fear, the Real Forensics Behind the Great Detective's Greatest Cases*, Hoboken, NJ: John Wiley & Sons, 2006.

2 L. Steenberg, *Forensic Science in Contemporary American Popular Culture: Gender, Crime, and Science*, New York and Abingdon: Routledge, 2013.

3 C. Strauss, 'The imaginary', *Anthropological Theory*, 2006, 6 (3): 322–344.
4 C. A. Mitchell, *Forensic Chemistry in the Criminal Courts*, London: The Institute of Chemistry of Great Britain and Northern Ireland, 1938; J. B. Firth, 'Forensic science laboratories', *Medico-Legal and Criminological Review*, 1945, 13 (3): 120–133.
5 For crises in masculinity, see J. A. Kestner, *Sherlock's Men: Masculinity, Conan Doyle and Cultural History*, Aldershot and Brookfield USA: Ashgate, 1997. For anxieties about loss of empire see J. A. Kestner, *The Edwardian Detective, 1901–1915*, Aldershot and Brookfield USA: Ashgate, 2000.
6 J. K. Van Dover, *You Know My Method: The Science of the Detective*, Bowling Green, OH: Bowling Green State University Popular Press, 1994.
7 R. R. Thomas, *Detective Fiction.*
8 N. Donaldson, *In Search of Dr Thorndyke: The Story of R. Austin Freeman's Great Scientific Investigator and His Creato*r, Bowling Green, OH: Bowling Green University Popular Press, 1971.
9 Van Dover, *You Know My Method.*
10 See A. B. Reeve, *The Silent Bullet: The Early Exploits of Craig Kennedy, Scientific Detective*. London: Hodder and Stoughton, 1916 (originally published: New York: Dodd, Mead, 1912); J. Locke, *From Ghouls to Gangsters: The Career of Arthur B. Reeve*, Elkhorn CA: Off-Trail Publications, 2007, 2 vols.
11 A. Conan Doyle, 'A Study in Scarlet', originally published 1887, Available at http://www.gutenberg.org/files/244/244-h/244-h.htm, accessed 29 April 2015.
12 D.S.L. Cardwell, *The Organisation of Science in England: A Retrospect*, Melbourne, London, Toronto: Wm Heinemann Ltd, 1957.
13 S. Shapin, *The Scientific Life: A Moral History of a Late Modern Vocation*, Chicago, IL and London: University of Chicago Press, 2008, p. 33.
14 R. Haynes, *From Faust to Strangelove: Representations of the Scientist in Western Literature*, Baltimore, MD: The Johns Hopkins University Press, 1994; R. Haynes, 'From alchemy to artificial intelligence: Stereotypes of the scientist in Western literature', *Public Understanding of Science*, 2003, 12 (3): 243–253; N. Russell,' Science and scientists in Victorian and Edwardian literary novels: Insights into the emergence of a new profession', *Public Understanding of Science*, 2007, 16 (2): 205–222; N. Russell, 'The new men: Scientists at work in popular British fiction between the early 1930s and the late 1960s,' *Science Communication*, 2009, 31 (1): 29–56.
15 R. Haynes, 'From alchemy'.
16 N. Russell, 'Science and scientists'.
17 G. Gissing, *Born in Exile*, London: Victor Gollancz, 1970 (originally published 1892).
18 L. Nocks, 'T. H. Huxley: The evolution of the bulldog', in B. Regal (ed), *Icons of Evolution: An Encyclopedia of People, Evidence, and Controversies*, Volume 1, Westport, CT: ABC-CLIO/Greenwood, 2008, 57–86, p. 76.
19 R. A. Freeman, *The Red Thumb Mark*, Kelly Bray, Cornwall, House of Stratus, 2001, p. 4, (originally published 1907).
20 Conan Doyle, *A Study in Scarlet.*
21 J. B. Firth,. *A Scientist Turns to Crime*, London: William Kimber, 1960, pp. 16–17.
22 'Dr Julius Gant: Obituary' The Telegraph, 8 July 1991, available at http://www.telegraph.co.uk/news/obituaries/7859337/Dr-Julius-Grant.html, accessed 17 April 2015.
23 H. E. Cox, 'Charles Ainsworth Mitchell (1867–1948)', *The Analyst*, 1948, 73: 55–57.
24 Haynes, *From Faust to Strangelove*; Haynes, 'From alchemy', p. 244.
25 Haynes, *From Faust to Strangelove.*
26 Rzepka describes the hard-boiled genre thus: '"tough-guy" crime fiction was conceived in part as a direct challenge to the Anglo-American classical tradition inspired

by Holmes'. C. J. Rzepka, *Detective Fiction*, Cambridge and Malden, MA: Polity, 2005, p. 179.

27 J. Verne, *A Journey to the Centre of the Earth*, London: Griffith and Farran, 1872.
28 Haynes, 'From alchemy', p. 250.
29 Van Dover, *You Know My Method*, pp. 134–136.
30 Donaldson, *In Search of Dr Thorndyke*, pp. 125–130.
31 Van Dover, *You Know My Method*, pp. 134–136.
32 H. Shpayer-Makov, 'Shedding the uniform and acquiring a new masculine image: The case of the late-Victorian and Edwardian police detective', in D. G. Barrie and S. Broomhall (eds), *A History of Police and Masculinities, 1700–2010*, London and New York: Routledge, 2012, 141–162, p. 149.
33 H. J. Walls, *Expert Witness: My Thirty Years in Forensic Science*, London: John Long, 1972.
34 A. Conan Doyle, *The Sign of Four*, 1890, available at http://www.gutenberg.org/files/2097/2097-h/2097-h.htm, Chapter 1 The Science of Deduction, accessed 29 April 2015.
35 A. Conan Doyle, 'The Final Problem',1893, *Memoirs of Sherlock Holmes*, London: George Newnes, Reprinted in A. Conan Doyle *The Treasury of Sherlock Holmes, 7 – in 1 Omnibus edition*, Radford VA: Wilder, 2007, 291-298, p. 292.
36 Ibid.
37 A. Conan Doyle, 'The Boscombe Valley Mystery', available at The Adventures of Sherlock Holmes 1891, http://www.gutenberg.org/files/1661/1661-h/1661-h.htm, accessed 29 April 2015.
38 E. A. Poe, 'The murders in the Rue Morgue', *Graham's Magazine*, April 1841, 166–179; W. Collins, *The Moonstsone*, London: Tinsley Brothers, 1868.
39 For the *Baker Street Journal* see http://www.bakerstreetjournal.com/.
40 Conan Doyle, 'The Boscombe Valley Mystery'; J. Adam and J. C. Adam, *Criminal Investigation A Practical Textbook for Magistrates, Police Officers and Lawyers*, Madras: A. Krishnamachari, 1906; A. Lucas, *Forensic Chemistry*, London: Edward Arnold, 1921.
41 A. Conan Doyle, *Memories and Adventures*, London: Hodder and Stoughton, 1924, p. 112
42 Ibid., p. 72.
43 J. F. O'Brien, *The Scientific Sherlock Holmes: Cracking the Case with Science and Forensics*, Oxford and New York: Oxford University Press, 2013, p. 154.
44 K. Simpson, *Sherlock Holmes on Medicine and Science*, With an Introduction by Isaac Asimov BSI and An Appreciation by E. Stanley Palm, New York: Magico Magazine, 1983.
45 A. Conan Doyle, *The Hound of the Baskervilles*, London: George Newnes, 1902, available at https://www.gutenberg.org/files/2852/2852-h/2852-h.htm, accessed 30 April 2015; J. Tyndall, *Scientific Use of the Imagination: And Other Essays*, London: Longmans, Green, and Co., 1872.
46 S. Smith, *Mostly Murder*, London: Harrap, 1959, Firth; *A Scientist Turns to Crime*, Simpson, *Sherlock Holmes*; and I. Asimov, 'Introduction' in K. Simpson, *Sherlock Holmes on Medicine and Science*, With an Introduction by Isaac Asimov BSI and An Appreciation by E. Stanley Palm, New York: Magico Magazine, 1983, vii-viii.
47 S. O Berg, 'Sherlock Holmes: Father of scientific crime detection' *Journal of Criminal Law and Criminology*, 1970, 61 (3): 446–452.
48 Van Dover, *You Know My Method*, p. 129.
49 Ibid., p. 133.
50 Freeman, *The Red Thumb Mark*, p. 5.

51 Van Dover, *You Know My Method*, p. 143.
52 Ibid., p. 135.
53 Ibid., p. 136, quoting Freeman, 'Meet Dr Thorndyke', p. xiv.
54 Van Dover, *You Know My Method*, pp. 140–141.
55 Thorndyke would not have called himself a forensic scientist, as the term was not in common use until the 1930s and his first appearance was in 1907 in R. A. Freeman, *The Red Thumb Mark*, Kelly Bray, Cornwall, House of Stratus, 2001 (Original edition London: Collingwood, 1907). Recent UK editions of Thorndyke novels are published by House of Stratus. For TV and radio: R. A. Freeman, Mr Pottermack's Oversight, Broadcast on BBC Home Service, Saturday Night Theatre, 14 September 1963. http://genome.ch.bbc.co.uk/f511a92830254735807d9d52aceb726d, accessed 29 April, 2015; R. A. Freeman – Thorndyke, Forensic Investigator, series of readings of Thorndyke stories broadcast on BBC Radio 4 Extra in 2011 and 2013 (Note that the epithet 'forensic' appears which is not an adjective used in the original works), http://www.bbc.co.uk/programmes/b01r766x/episodes/guide, A dramatization of The Moabite Cipher, broadcast on BBC Radio 4, 21 January, 2015 as part of The Rivals, a series of Sherlock Holmes Rivals, http://www.bbc.co.uk/programmes/b04ykbjw. For 1964 BBC TV series see: http://www.startrader.co.uk/Action%20TV/guide60s/thorndyke.htm. All accessed 29 April 2015.
56 Donaldson, *In Search of Dr Thorndyke*; and R. A. Freeman, 'Meet Dr Thorndyke', in R. A. Freeman, *Dr Thorndyke's Crime File*, P. M. Stone (ed), New York: Dodd, Mead, vii–xv.
57 Freeman, *The Red Thumb Mark*, 1907; R. A Freeman, *The Jacob Street Mystery*, London: Hodder and Stoughton, 1942.
58 R.A. Freeman, *The Mystery of Angelina Frood*, London: Hodder and Stoughton, 1924. Reprinted in *Dr Thorndyke's Crime File*, New York: Dodd, Mead (ed. P.M. Stone), available at http://gutenberg.net.au/ebooks10/1000171.txt, accessed 12 May 2015. The identity of the mineral in this story is an example of the kind of specialist knowledge that Freeman's readers would not have known..
59 R. A. Freeman, *The Shadow of the Wolf*, Guildford and Esher: Billing and Sons Ltd, 1925.
60 Donaldson, *In Search of Dr Thorndyke*, p. 212.
61 Ibid., p. 114. R. A. Freeman, *A Silent Witness*, London: Hodder and Stoughton, 1914.
62 Freeman, 'Meet Dr Thorndyke'.
63 Ibid., p. viii.
64 For Bell's influence, see Conan Doyle, *Memories and Adventures*, pp. 25–26. For Thorndyke modelled on Taylor, see Donaldson, *In Search of Dr Thorndyke*, p. 66.
65 Freeman, 'Meet Dr Thorndyke', p. vii.
66 Ibid., p. xiv.
67 Van Dover, *You Know My Method*, p. 146.
68 Ibid., p. 149.
69 A. Conan Doyle, 'The naval treaty' in *The Memoirs of Sherlock Holmes*, London: George Newnes, 1894, available at http://www.gutenberg.org/files/834/834-h/834-h.htm#link2H_4_0010, accessed 30 April, 2015.
70 A. Conan Doyle, *The Hound of the Baskervilles*, London: George Newnes, 1902. https://www.gutenberg.org/files/2852/2852-h/2852-h.htm, accessed 30 April, 2015, Chapter 1, Mr. Sherlock Holmes.
71 O'Brien, *The Scientific Sherlock Holmes*, p. 50.

72 F. Galton, *Finger Prints*, London and New York: Macmillan, 1892. See S. A. Cole, *Suspect Identities: A History of Fingerprinting and Criminal Identification*, Cambridge, MA: Harvard University Press, 2001, for a history of fingerprinting.
73 M. Twain, *Pudd'nhead Wilson*, London: Chatto & Windus, 1894.
74 O'Brien, *The Scientific Sherlock Holmes*, p. 53.
75 A. Conan Doyle, 'The adventure of the Norwood builder', *The Return of Sherlock Holmes*, London: George Newnes, 1905, available at http://www.gutenberg.org/files/108/108-h/108-h.htm#linkH2H_4_0002, accessed 29 April 2015.
76 O'Brien, *The Scientific Sherlock Holmes*, p. 53, R. A. Freeman, *The Red Thumb Mark.*
77 There is a suggestion that Conan Doyle got the idea of using a forged fingerprint from the *Daily Express* journalist Bertram Fletcher Robinson on a sea voyage in 1901 where they were on the same boat. Conan Doyle is said to have paid him £50 for the idea. See 'The Hound of the Baskervilles Part 1' Available at http://www.bfronline.biz/index.php?option=com_content&task=view&id=47&Itemid=9, accessed 20 April 2015.
78 A. Conan Doyle, 'The adventure of the Norwood builder'.
79 Freeman, *The Red Thumb Mark*, p. i.
80 Ibid.
81 Richards and Gore, *Holmes, Chemistry.*
82 A. J. Richards and B. Gore, 'Sherlock Holmes: A study in chemistry', in A. J. Richards, and B. Gore, *Holmes, Chemistry and the Royal Institution*, London: Sherlock Holmes Society of London, 1998, 23–34.
83 Ibid., pp. 29–30.
84 Ibid.
85 Lindsey, *Sherlock Holmes and a Question of Science.*
86 Ibid., p. 98.
87 Ibid., emphasis is in the original text.
88 Wagner, *The Science of Sherlock Holmes.*
89 O'Brien, *The Scientific Sherlock Holmes.*
90 Ibid., p. 50.
91 Ibid., pp. 55–60.
92 O'Brien, *The Scientific Sherlock Holmes*, p. 117; Lindsey, *Sherlock Holmes and a Question of Science*, p.135, A. Conan Doyle, 'The adventure of Shoscombe Old Place' (1927), reprinted in *The Second Sherlock Holmes Illustrated Omnibus*, London: John Murray/Jonathan Cape, 1979, 377–388.
93 O'Brien, *The Scientific Sherlock Holmes*, p. 117. For Isaac Asimov's views, see I. Asimov, 'Introduction'.
94 Donaldson, *In Search of Dr Thorndyke*, pp. 91–92.
95 Cf Freeman, *The Red Thumb Mark*, pp. 181–204; Conan Doyle 'The Norwood builder'.
96 Firth,. *A Scientist Turns to Crime*, p. 13.
97 Simpson, *Sherlock Holmes on Medicine and Science*, p. 10.
98 Ibid., p. 11.
99 C. A. Mitchell, *The Scientific Detective and the Expert Witness*, Cambridge: W. Heffer & Sons Ltd, 1931, p. 2.
100 M. Gilberg, 'Alfred Lucas'.
101 Steenberg, *Forensic Science*, 25–26.
102 Ibid., p. 107.

103 See 'Portillo's State Secrets' No. 1 Crime and Punishment. First broadcast on BBC2 on 23 March 2015. Details at http://www.bbc.co.uk/programmes/b05p6pkc, accessed on 20 April 2015.
104 A. Conan Doyle, *Memories and Adventures*, London: Hodder and Stoughton, 1924, p. 108.
105 Details at http://www.sherlock-holmes.co.uk, accessed on 20 April, 2015.
106 R. Oldfield, *Outrage: The Edalji Five and the Shadow of Sherlock Holmes*, Cambridge: Vanguard Press, 2010, p. 91.
107 Ibid., pp.101–102.
108 Ibid., p. 92, *Daily Telegraph*, 11 January 1907.
109 S. O. Berg, 'Sherlock Holmes: Father of scientific crime detection', *Journal of Criminal Law and Criminology*, 1970, 61 (3): 446–452, p. 446.
110 Ibid., p. 448.; I. Burney, 'Our environment in miniature: Dust and the early twentieth-century forensic imagination', *Representations*, 2013, 121 (1): 31–59, p. 33.
111 M. Mazévet, *Edmond Locard: Le Sherlock Holmes Français*, Lyon: Editions des Traboule, 2006; M. Gilberg, 'Alfred Lucas: Egypt's Sherlock Holmes', *Journal of the American Institute for Conservation*, 1997, 36(1): 31–48.
112 D. Gardiner and K. Sorley Walker (eds.) (1997) *Raymond Chandler Speaking*, Berkeley and Los Angeles: University of California Press, pp. 59–60.
113 Ibid., p. 60.
114 Gilberg, 'Alfred Lucas'.
115 R. A. Freeman, *The Mystery of Angelina Frood*, London: Hodder and Stoughton, 1924. Reprinted in *Dr Thorndyke's Crime File*, New York: Dodd, Mead (ed. P. M. Stone), available at http://gutenberg.net.au/ebooks10/1000171.txt, accessed 12 May 2015.
116 J. Goodman, *The Crippen File*, London and New York: Allison & Busby: 1985. p. 89, Letter from H. G. Stevenson Coppin to Sir William Wilcox, 5 January 1925 and response from Sir William Wilcox to H. G. Stevenson Coppin, 9 January 1925.
117 Freeman, *Angelina Frood*, Chapter XXVII (reference is to online edition, hence no page numbers).
118 Ibid.
119 Ibid.
120 Ibid.
121 Donaldson, *In Search of Dr Thorndyke*, p. 152.
122 Goodman, *The Crippen File*, p. 89.
123 J. Glaister Jnr., *Final Diagnosis*, London: Hutchinson, 1964, p. 189.
124 Ibid.
125 BBC TV series *The Expert*, available at http://www.screenonline.org.uk/tv/id/481801/index.html, accessed 12 May 2015.
126 I. Burney, 'Our environment', p. 53.
127 Ibid.; J. C. Adam, *Criminal Investigation A Practical Textbook for Magistrates, Police Officers and Lawyers*, London: Sweet and Maxwell, 1924, p. 102.
128 N. Kendal, *Criminal Investigation A Practical Textbook for Magistrates, Police Officers and Lawyers*, London: Sweet and Maxwell, 1934, 3rd edition.
129 A. Pentland, 'Detective's portable forensic and finger-print outfit', *The Police Journal*, 1934, 7 (1): 108–110.

Bibliography

Newspapers

Daily Mail
Manchester Guardian
News of the World
New York Times
Telegraph

Newsreels

British Pathé, *Burglars Beware* (1935), available at http://www.britishpathe.com/video/burglers-beware-aka-burglars-beware/query/Hendon+laboratory, accessed 6 May 2015.

British Pathé, *Science Fights Crime* (1946), available at http://www.britishpathe.com/video/science-fights-crime/query/for, accessed 5 May 2015.

Websites, radio and TV broadcasts

Baker Street Journal, available at http://www.bakerstreetjournal.com/, accessed 28 April, 2015.

BBC TV series *The Expert*, available at http://www.screenonline.org.uk/tv/id/481801/index.html, accessed 12 May 2015.

Finn, C. 'Recreating the sound of Tutankhamun's trumpet', available at http://www.bbc.co.uk/news/world-middle-east-13092827, accessed 11 May 2015.

Freeman, R.A., Mr Pottermack's Oversight, Broadcast on BBC Home Service, Saturday Night Theatre, 14 September 1963, available at http://genome.ch.bbc.co.uk/f511a92830254735807d9d52aceb726d

—— Thorndyke, Forensic Investigator, series of readings of Thorndyke stories broadcast on BBC Radio 4 Extra in 2011 and 2013, available at http://www.bbc.co.uk/programmes/b01r766x/episodes/guide

—— The Moabite Cipher, broadcast on BBC Radio 4, 21 January, 2015, part of The Rivals, a series of Sherlock Holmes Rivals, available at http://www.bbc.co.uk/programmes/b04ykbjw, accessed 29 April, 2015.

—— Dr Thorndyke 1964 BBC TV series, http://www.startrader.co.uk/Action%20TV/guide60s/thorndyke.htm, accessed 29 April, 2015.

'Pereira, Margaret', *Who's Who 2014*, A & C Black, an imprint of Bloomsbury Publishing plc, 2014; Oxford University Press, 2014; online edn, Nov 2014, available at http://www.ukwhoswho.com/view/article/oupww/whoswho/U30563, accessed 2 May 2015.

'Portillo's State Secrets' No. 1 Crime and Punishment. First broadcast on BBC2 on 23 March 2015, available at http://www.bbc.co.uk/programmes/b05p6pkc, accessed on 20 April 2015.

Robinson, B. F, 'The Hound of the Baskervilles Part 1', available at http://www.bfronline.biz/index.php?option=com_content&task=view&id=47&Itemid=9, accessed 20 April 2015.

Primary sources, fiction, contemporary books, articles, biographies and obituaries

Adam, J. C., *Criminal Investigation A Practical Textbook for Magistrates, Police Officers and Lawyers*, Translated and adapted from the System der Kriminalistik of Dr Hans Gross, London: Sweet and Maxwell, 1924.

Adam, J. and Adam, J. C., *Criminal Investigation A Practical Handbook for Magistrates, Police Officers and Lawyers*, Madras: A. Krishnamachari, 1906.

—— *Criminal Investigation A Practical Handbook for Magistrates, Police Officers and Lawyers*, London: The Specialist Press, 1907 (reprint of 1906 edition).

Anon., 'The opening ceremony of the Metropolitan Police Laboratory, *Metropolitan Police College Journal*, 1935, 1: 19–21.

Anon., 'Julius Grant 1901–1991', *Journal of the Forensic Science Society*, 1991, 31 (3): 397–399.

Asimov, I., 'Introduction' in K. Simpson, *Sherlock Holmes on Medicine and Science*, With an Introduction by Isaac Asimov BSI and An Appreciation by E. Stanley Palm, New York: Magico Magazine, 1983, vii–viii.

Beccaria, C., *On Crimes and Punishments* (1764) (English translation 1767), 1767, available at https://archive.org/details/anessayoncrimes00beccgoog, accessed 24 January 2013.

Belper Committee, *Method of Identification of Criminals by Measurements and Finger Prints*, London: Printed for Her Majesty's Stationery Office by Wyman and Sons Ltd., 1901.

Bentham, J., *A Fragment on Government*, Payne: London, 1776. Available at http://www.constitution.org/jb/frag_gov.htm, accessed 12 January 2015.

Bertillon, A., *L'identité des Récidivistes et la Loi de Relégation*, Paris: G. Masson, 1883.

Booth-Tucker, F., *Muktifauj, or, Forty Years with the Salvation Army in India and Ceylon*, London and Edinburgh: Marshall Brothers, 1923.

Camps, F. E., *Camps on Crime*, Newton Abbot: David & Charles, 1973.

Cherill, F. R., *The Fingerprint System of Scotland Yard: A Practical Treatise on Finger Print Identification for the Use of Students and Experts and A Guide for Investigators when Dealing with Imprints Left at the Scenes of Crime*, London: HMSO: 1954.

Coleman, R. F. and Walls, H. J., 'The evaluation of scientific evidence', *The Criminal Law Review*, 1974: 276–288.

Collins, W., *The Moonstone*, London: Tinsley Brothers, 1868.

Conan Doyle, A. *A Study in Scarlet*, originally published 1887, available at http://www.gutenberg.org/files/244/244-h/244-h.htm, accessed 28 April 2015.

—— *The Sign of Four*, 1890, available at http://www.gutenberg.org/files/2097/2097-h/2097-h.htm, accessed 29 April 2015.

—— 'The Boscombe Valley mystery', in *The Adventures of Sherlock Holmes*, 1892, available at http://www.gutenberg.org/files/1661/1661-h/1661-h.htm, accessed 28 April 2015.

—— 'The Red Headed League', in *The Adventures of Sherlock Holmes*, George Newnes Ltd, 1892, available at http://www.gutenberg.org/files/1661/1661-h/1661-h.htm, accessed 28 April 2015.

—— 'The Final Problem', 1893, *Memoirs of Sherlock Holmes*, London: George Newnes, Reprinted in A. Conan Doyle *The Treasury of Sherlock Holmes, 7 – in 1 Omnibus edition*, Radford VA: Wilder, 2007, 291–298.

—— 'The naval treaty' in *The Memoirs of Sherlock Holmes*, London: George Newnes, 1894, available at http://www.gutenberg.org/files/834/834-h/834-h.htm#link2H_4_0010, accessed 30 April 2015.

—— *The Hound of the Baskervilles*, London: George Newnes, 1902, available at https://www.gutenberg.org/files/2852/2852-h/2852-h.htm, accessed 30 April 2015.

—— *Memories and Adventures*, London: Hodder and Stoughton, 1924.

—— 'The adventure of the Norwood builder', in *The Return of Sherlock Holmes*, London: George Newnes, 1905, available at http://www.gutenberg.org/files/108/108-h/108-h.htm#linkH2H_4_0002, accessed 29 April 2015.

—— *The Adventure of Shoscombe Old Place* (1927), reprinted in *The Second Sherlock Holmes Illustrated Omnibus*, London: John Murray/Jonathan Cape, 1979, 377–388.

Cox, H. E., 'Charles Ainsworth Mitchell (1867–1948)', *The Analyst*, 1948, 73: 55–57.

Crawley, F. J., 'Decentralization and the police box system', *The Police Journal*, 1928, 1(1): 118–127.

Cuthbert, C.R.M., *Science and the Detection of Crime*, London: Hutchinson, 1958.

Devon, J., *The Criminal and the Community*, London: John Lane/Bodley Head, 1912.

Dew, W. J., *I Caught Crippen: Memoirs of Ex-Chief Inspector Walter Dew, C.I.D. of Scotland Yard*, London and Glasgow: Blackie & Son Ltd, 1938.

Dickens, C., *Hard Times*, Harmondsworth: Penguin, 1969 (originally published 1854).

Dixon, A. L., 'The English police system', *Annals of the American Academy of Political and Social Sciences*, 1929, 146: 177–192.

—— 'The new series of the Police Journal', *The Police Journal*, 1933, 6 (1): 1–3.

—— *Atomic Energy for the Layman*, London: Chantry Publications, 1950.

Dower, A., *Crime Chemist: The Life Story of Charles Anthony Taylor Scientist for the Crown*, London: J. Long, 1965.

Dunlap, A. 'Science versus practical common sense in crime detection', *American Journal of Police Science*, 1931, 2 (4): 322–327.

Dyer, B. and Mitchell, C.A., *The Society of Public Analysts and Other Analytical Chemists: Some Reminiscences of Its First Fifty Years and Review of Its Activities*, Cambridge: Heffer, 1932.

Ellis, H., *The Criminal*, New York: Scribner and Welford, 1890.

Else, W. M. and Garrow, J. M., *The Detection of Crime: An Introduction to Some Methods of Scientific Aid in Criminal Investigation*, London: The Police Journal, 1934.

Faulds, H., 'On the skin-furrows of the hand', *Nature*, 1880, 22 (October 28): 605.

Firth, J. B., 'Forensic science laboratories', *Medico-Legal and Criminological Review*, 1945, 13 (3): 120–133.

—— *A Scientist Turns to Crime*, London: William Kimber, 1960.

Freeman R.A., *The Red Thumb Mark*, Kelly Bray, Cornwall, House of Stratus, 2001 (Original edition London: Collingwood, 1907).

—— *A Silent Witness*, London: Hodder and Stoughton, 1914.

—— *The Mystery of Angelina Frood*, London: Hodder and Stoughton, 1924. Reprinted in *Dr Thorndyke's Crime File*, New, York: Dodd, Mead (ed. P. M. Stone), available at http://gutenberg.net.au/ebooks10/1000171.txt, accessed 12 May 2015.

—— *The Shadow of the Wolf*, Guildford and Esher: Billing and Sons Ltd, 1925.

—— 'Meet Dr Thorndyke', in R. A. Freeman, *Dr Thorndyke's Crime File*, New York: Dodd, Mead, vii–xv (ed. P. M. Stone), 1941, available at http://gutenberg.net.au/ebooks10/1000171.txt, accessed 20 April 2015.

—— *The Jacob Street Mystery*, London: Hodder and Stoughton, 1942.
Galton, F., *Hereditary Genius: An Inquiry into Its Laws and Consequences*. London: Macmillan, 1869.
—— 'Composite portraits', *Nature*, 1878, 18, May 23: 97–100.
—— *Generic Images*, London: William Clowes and Son, 1879.
—— *Inquiry into Human Faculty and Its Development*, London: Macmillan, 1883.
—— *Finger Prints*, London and New York: Macmillan, 1892.
Gissing, G., *Born in Exile*, London: Victor Gollancz, 1970 (originally published 1892).
Glaister Jnr., J., 'Some results of recent medico-legal research in the examination of bloodstains and hairs', *The Police Journal*, 1928, 1(1): 62–77.
—— *Final Diagnosis*, London: Hutchinson, 1964.
Goring, C., *The English Convict: A Statistical Study*, London: HMSO, 1913.
Grant, J., *Science for the Prosecution*, London: Chapman and Hall, 1941.
—— 'Presidential address: Forensic scientist at large', *Medico-Legal Journal*, 1973, 41 (4): 132–141.
—— 'The past, present and future role of the private forensic science laboratory', *Journal of the Forensic Science* Society, 1977, 16 (3): 197–200.
—— 'Forensic scientist – Still at large!', *Medico-Legal Journal*, June 1982, 50 (2): 61–74.
Grassberger, R., 'Pioneers in criminology XIII. Hans Gross (1847–1915)', *Journal of Criminal Law, Criminology, and Police Science*, 1956, 47 (4): 397–405.
Gross, H., *Handbuch für Untersuchungsrichter als System der Kriminalistik*, 2 volumes, Munich: J. Schweitzer Verlag, 1893.
—— *Criminal Psychology: A Manual for Judges, Practitioners, and Students by Hans Gross*, London: Heinemann 1911 (translated from the fourth German edition by Horace M. Kallen).
Hall-Caine, G., 'Banditry Bill', *Hansard*, 24 March 1933, available at http://hansard.millbanksystems.com/commons/1933/mar/24/banditry-bill, accessed 22 April 2015.
Hastings, M., *The Other Mr Churchill: A Lifetime of Shooting and Murder*, London: George Harrap & Co. Ltd, 1963.
Herschel, W. J. *The Origin of Finger-Printing*, London: Humphrey Milford/ Oxford University Press, 1916.
HMSO, *Home Office: Report of the Advisory Committee on the Scientific Investigation of Crime*, London: His Majesty's Stationery Office, 1936.
—— *Home Office: Scientific Aids to Criminal Investigation, Forensic Science Circulars*, No. 1, HMSO, London, 1936.
—— *Home Office: Scientific Aids to Criminal Investigation: Instructional Pamphlet for the Use of Police Officers*, His Majesty's Stationery Office, 1936 (subsequent editions: 1949/1950, reprinted 1964).
—— *Home Office: Report of the Departmental Committee on Detective Work and Procedure*, London: His Majesty's Stationery Office, 1938.
Hogan, J. C. and Schwartz, M. D., 'The manly art of observation and deduction', *Journal of Criminal Law and Criminology*, 1964, 55 (1): 157–164.
Howe, R. M., *Criminal Investigation A Practical Textbook for Magistrates, Police Officers and Lawyers*. London: Sweet and Maxwell, 1949, 4th edition.
Huxley, T. H., *Evidence as to Man's Place in Nature*, London: Williams & Norgate, 1863.
Jackson, R. L., *Criminal Investigation A Practical Textbook for Magistrates, Police Officers and Lawyers*, London: Sweet and Maxwell, 1962, 5th edition.
Jago, W., *The Chemistry of Wheat, Flour, Bread and Technology of Breadmaking*, Brighton: William Jago, 1886.

—— *A Manual of Forensic Chemistry Dealing Especially with Chemical Evidence, Its Preparation and Adduction. Based upon a Course of Lectures delivered at University College, University of London*, London: Stevens and Haynes, 1909.

Jago, W. and Jago, W. C., *The Technology of Bread-Making: Including the Chemistry and Analytical and Practical Testing of Wheat, Flour, and Other Materials Employed in Bread-making and Confectionery*, London: Simpkin, Marshall, Hamilton, Kent & Co., 1911.

Kendal, N., *Criminal Investigation A Practical Textbook for Magistrates, Police Officers and Lawyers*, London: Sweet and Maxwell, 1934, 3rd edition.

Kent-Jones, D. W., 'Obituary notices: William Jago 1853–1938', *Journal of the Chemical Society*, 1938: 1127–1128.

Kopp, I. *The Fantastic Life of Harry Söderman 1902–1956*, National Laboratory of Forensic Science – SKL/ National Forensic Centre, available at http://nfc.polisen.se/Global/www%20och%20Intrapolis/Informationsmaterial/SKL/Soderman_minnesskrift.pdf, 2015, accessed 21 April 2015.

Locard, E. 'Dust and its analysis: An aid to criminal investigation', *The Police Journal*, 1928, 1 (2): 177–192.

Lombroso, C., *Criminal Man*, Translated by M. Gibson and N. H. Rafter, Durham, NC: Duke University Press, 2006 (original publication date 1876).

Lucas, A., *Ministry of Finance, Egypt, Report on the Work of the Government Analytical Laboratory and Assay Office During the Period 1913–19 by A. Lucas, F.I.C., Director*. Cairo: Government Press, 1920.

—— *Legal Chemistry and Scientific Criminal Investigation*, London: Edward Arnold, 1920.

—— *Forensic Chemistry*, London: Edward Arnold, 1921.

—— 'Government Analytical Laboratory, Cairo, Report of the Director for the Year 1920', D. R. Wood and R. B. Pilcher, 'Notes from the Reports of Public Analysts', *The Analyst*, 1922, 47 (19–24): 21–23.

—— *Forensic Chemistry and Scientific Criminal Investigation*, London: Edward Arnold, 1931, 2nd edn.

—— *Forensic Chemistry and Scientific Criminal Investigation*, London: Edward Arnold, 1935, 3rd edn.

—— *Forensic Chemistry and Scientific Criminal Investigation*, London: Edward Arnold, 1945, 4th edn.

—— *Ancient Egyptian Materials*, London: Edward Arnold, 1926.

Maxwell, J., 'The English police system: General developments and outstanding features', in L. Radzinowicz and J.W.C. Turner (eds), *Penal Reform in England*, London: Macmillan, 1946, 60–97.

McCafferty, J., *Mac, I've Got a Murder*, London: Arthur Barker Ltd., 1975.

Mitchell, C. A., *Science and the Criminal*, London: Isaac Pitman & Sons, 1911.

—— *Documents and Their Scientific Examination: With Especial Reference to the Chemistry Involved in Cases of Suspected Forgery, Investigation of Disputed Documents, Handwriting, etc.*, London: C. Griffin, 1922.

—— 'Obituary. Otto Hehner', *The Analyst*, 1924, 49 (584): 501–505.

—— 'Science as applied to circumstantial evidence', *The Police Journal*, 1928, 1 (2): 256–267.

—— *The Scientific Detective and the Expert Witness*, Cambridge: W. Heffer & Sons Ltd, 1931.

—— *Forensic Chemistry in the Criminal Courts*, London: The Institute of Chemistry of Great Britain and Northern Ireland, 1938.

—— 'Review of *Science for the Prosecution*', *The Analyst*, 1942, 67: 154.
—— *A Scientist in the Criminal Courts*, London: Chapman & Hall Ltd, 1945.
Morland, N. *Fingerprints: An Introduction to Scientific Criminology*, London: Street & Massey, 1936.
—— *The Conquest of Crime*, London: Cassell and Co., 1937.
—— *An Outline of Scientific Criminology*, London: Cassell and Co., 1950.
—— *Papers from the Criminologist*, London: Wolf Publishing Ltd., 1971.
—— 'Review: *Science Against Crime*', *Journal of the Forensic Science Society*, 1974, 14 (1): 77–78.
Morrish, R., *The Police and Crime-Detection To-Day*, London: Oxford University Press, 1946.
Nickolls, L. C., *The Scientific Investigation of Crime*, London: Butterworth, 1956.
Osborn, A. S., *Questioned Documents: A Study of Questioned Documents with an Outline of Methods by which the Facts May be Discovered and Shown*, Rochester, NY: The Lawyers' Co-operative Publishing Co., 1910.
Pentland, A., 'Detective's portable forensic and finger-print outfit', *The Police Journal*, 1934, 7 (1): 108–110.
Pereira, M., 'Dr Henry James Walls', *Journal of the Forensic Science Society*, 28 (5–6), 1988: 357–358.
Pilcher, R. B., *A List of Official Chemical Appointments in Great Britain and Ireland, in India and the Colonies*, London: Institute of Chemistry, 1906.
Poe, E. A., 'The murders in the Rue Morgue', *Graham's Magazine*, April 1841: 166–179.
Police Journal, 'Editorial', *The Police Journal*, 1928, 1 (1): 1–2.
Potter, S., *The Theory and Practice of Gamesmanship: Or the Art of Winning Games Without Actually Cheating*, London: Rupert Hart-Davis, 1947.
—— *One-Upmanship. . . Being Some Account of the Activities and Teaching of the Lifemanship Correspondence College of One-Upness and Gameslifemastery*, London: Rupert Hart-Davis, 1952.
Radley, J. A. and Grant, J., *Fluorescence Analysis and Ultra-Violet Light*, London: Chapman and Hall, 1935 (4th edition 1954).
Reeve, A. B., *The Silent Bullet: The Early Exploits of Craig Kennedy, Scientific Detective*, London: Hodder and Stoughton, 1916 (originally published: New York : Dodd, Mead, 1912).
Reiss, R. A., *Manuel de Police Scientifique (Technique)*: *I. Vols et Homicides*, Lausanne: Payot, 1911.
Rhodes, H.T.F., 'The nature of chemical evidence and its place in the detection of crime', *The Police Journal*, 1928, 1(4): 657–668.
—— *Alphonse Bertillon: Father of Scientific Detection*, London: George G. Harrap & Co., 1956.
Roy, W. 'Traffic control in Edinburgh', *The Police Journal*, 1930, 3 (2): 262–276.
Scaplehorn, A., 'Obituary: Ian George Holden BSc PhD AKC FRIC, 1922–2011', *Science and Justice*, 51: 216.
Simpson, K., *Taylor's Principles and Practice of Medical Jurisprudence*, London: J. & A. Churchill, 1965, 2 vols., 12th edition.
—— 'Foreword', N. Morland, *Papers from the Criminologist*. London: Wolf Publishing Ltd., 1971, 11–13.
—— 'Foreword', H. J. Walls, *Expert Witness: My Thirty Years in Forensic Science*, London: John Long, 1972, 11–12.
—— *Forty Years of Murder: An Autobiography*, London: Harrap, 1978.

—— *Sherlock Holmes on Medicine and Science*, With an Introduction by Isaac Asimov BSI and An Appreciation by E. Stanley Palm, New York: Magico Magazine, 1983.

Smith, R. *Technical/Scientific, Social and Industrial Expertise and Legal Frameworks 1983–84 (SERC/ESRC)*, Recorded interviews with leading practitioners and commentators on legal medicine and forensic science conducted by Dr Roger Smith, Dept. of History, University of Lancaster, Hamish J. Walls, 2 June 1983, M. Pereira, 23 June 1983, Raymond Williams 21 December, 1983, Wellcome Library, London, 1983–1984.

Smith, S., 'The Identification of Firearms and Projectiles: As Illustrated by the Case of the Murder of Sir Lee Stack Pasha', *The Police Journal*, 1928, 1 (3): 411–422.

—— *Mostly Murder*, London: Harrap, 1959.

Söderman, H., 'Science and criminal investigation', *Annals of the American Academy of Political and Social Science*, 1929, 146 (1): 237–248.

—— *Policeman's Lot: A Criminologists' Gallery of Friends and Felons*, London: Longmans, Green and Co., 1957.

Stapleton, D., 'Evidence of fractured glass in the investigation of crime', *Forensic Science Circular*; No. 6, London: HMSO, 1940.

Symons, C. T., 'Scientific aids in prospect', *Metropolitan Police College Journal*, 1935, 1 (2): 30–34.

Taylor, A. S., *The Principles and Practice of Medical Jurisprudence*, London: Churchill, 1865.

Taylor, F. W., *Scientific Management: Comprising Shop Management, the Principles of Scientific Management, Testimony Before the Special House Committee*, New York: Harper, 1911.

Thompson, J., *Crime Scientist*, London: Harrap, 1980.

Thorne Baker, T., *The Telegraphic Transmission of Photographs*, New York: Van Nostrand, 1910.

Troup Committee, *Identification of Habitual Criminals*, London: HMSO, 1894.

Tryhorn, F. G., 'The packing of exhibits', *The Police Journal*, 1935, 8 (1): 19–26.

—— 'The assessment of circumstantial evidence', *The Police Journal*, 1935, 8 (4): 401–411.

—— 'Scientific aids in criminal investigation Part I', *The Police Journal*, 1936, 9 (1): 33–41.

—— 'Scientific aids in criminal investigation. Part II Searching at the scene of crime', *The Police Journal*, 1936, 9 (2): 152–160.

—— 'Scientific aids in criminal investigation. Part III Searching and packing', *The Police Journal*, 1936, 9 (3): 303–317.

—— 'Scientific aids in criminal investigation. Part IV Dust', *The Police Journal*, 1936, 9 (4): 403–412.

—— 'The fracture of glass', *Forensic Science Circular*; 1936, No. 2, London: HMSO, 1–5.

Turfitt, G., 'The fracture of safety glass by revolver bullets'. *Forensic Science Circular*; No. 6, 1940, London: HMSO, 12–15.

Twain, M., *Pudd'nhead Wilson*, London: Chatto & Windus, 1894.

Tyndall, J., *Scientific Use of the Imagination: And Other Essays*, London: Longmans, Green, and Co., 1872.

Verne, J., *A Journey to the Centre of the Earth*, London: Griffith and Farran, 1872.

Voelcker, E. and Watson, R. S., 'Obituary: Henry Edward Cox', *The Analyst*, 1952, 77 (913): 169–170.

Walls, H. J., *Forensic Science*, London: Sweet and Maxwell, 1968.

—— *Expert Witness: My Thirty Years in Forensic Science*, London: John Long, 1972.

—— *Expert Witness: My Thirty Years in Forensic Science*, London: The Quality Book Club, 1973, 2nd edition.
—— 'The Forensic Science Service in Great Britain: A short history', *Journal of the Forensic Science Society*, 1976, 16: 273–277.
Willcox, P.H.A., *The Detective Physician: The Life and Work of Sir William Willcox 1870–1941*, London, Heinemann, 1970.
Willcox, W., 'Toxicology and crime', *The Police Journal*, 1928, 1 (1): 98–104.
Wilton, G. W., 'Review of Criminal Investigation by Hans Gross, 4th edition, Edited by Robert Howe,' *Journal of Criminal Law and Criminology*, 1952, 43(2): 281–282.

Secondary sources, doctoral theses and critical works

Adam, A., *Spontaneous Generation in the 1870s: Victorian Scientific Naturalism and its Relationship to Medicine*, unpublished PhD thesis, CNAA, Sheffield City Polytechnic, 1989.
—— *Artificial Knowing: Gender and the Thinking Machine*, London and New York: Routledge, 1998.
—— *Gender, Ethics and Information Technology*, Houndmills Basingstoke: Palgrave Macmillan, 2004.
Akrich, M., 'The de-scription of technical objects', in W. E. Bijker, and J. Law (eds), *Shaping Technology/Building Society: Studies in Sociotechnical Change*, Cambridge, MA: MIT Press, 1992, 205–224.
Ambage, N., *The Origins and Development of the Home Office Forensic Science Service, 1931–1967*, unpublished PhD thesis, Lancaster University, 1987.
Ambage, N. and Clark, M., 'Unbuilt Bloomsbury: Medico-legal institutes and forensic science laboratories in England between the wars', in M. Clark and C. Crawford (eds), *Legal Medicine in History*, Cambridge: Cambridge University Press, 1994, 293–313.
Arapostathis, S. and Gooday, G., *Patently Contestable: Electrical Technologies and Inventor Identities on Trial in Britain*, Cambridge, MA and London: MIT Press, 2013.
Arnold, D. 'Crime and crime control in Madras, 1858–1947', in A. A. Young, *Crime and Criminality in British India*, Tucson: University of Arizona Press, 62–88.
Aronson, J.D., *Genetic Witness: Science, Law, and Controversy in the Making of DNA Profiling*, New Brunswick, NJ: Rutgers University Press, 2007.
Arora, R. K. and Goyal, R., *Indian Public Administration*, New Delhi: Wishwa Prakashan, 2011.
Asdal, K., Brenna, B. and Moser, I. B. (eds), *Technoscience: The Politics of Interventions*, Oslo: Unipub, Oslo Academic Press, 2007.
Astrahan, M. M. and Chamberlin, D. D., 'Implementation of a structured English query language', *Communications of the ACM*, 1975, 18 (10): 580–588.
Atkins, P., *Liquid Materialities: A History of Milk, Science and the Law* Farnham: Ashgate, 2010.
Beauregard, E. and Bouchard, M., 'Cleaning up your act: Forensic awareness as a detection avoidance strategy', *Journal of Criminal Justice*, 2010, 38 (6): 1160–1166.
Becker, P., 'The criminologists' gaze at the underworld: Toward an archaeology of criminal writing', in P. Becker and R. F. Wetzell (eds), *Criminals and Their Scientists: The History of Criminology in International Perspective*, Cambridge and New York: Cambridge University Press, 2006, 105–133.

Becker, P. and Wetzell, R.F. (eds) *Criminals and Their Scientists: The History of Criminology in International Perspective*, Cambridge and New York: Cambridge University Press, 2006.

Bell, A. 'The development of forensic pathology in London, England: Keith Simpson and the Dobkin case, 1942', *Canadian Bulletin of Medical History/ Bulletin Canadien D'histoire de la Médecine*, 2012, 29 (2): 265–282.

Bench-Capon, T.J.M. (ed.), *Knowledge-Based Systems and Legal Applications*, London: Academic Press, 1991.

Berg, S. O., 'Sherlock Holmes: Father of scientific crime detection', *Journal of Criminal Law and Criminology*, 1970, 61 (3): 446–452.

Berridge, V., 'Health and medicine', in F.M.L. Thompson (ed.), *The Cambridge Social History of Britain 1750–1950: Volume 3: Social Agencies and Institutions*, Cambridge: Cambridge University Press, 1990, 171–242.

Bijker, W. E., *Of Bicycles, Bakelites, and Bulbs: Toward a Theory of Sociotechnical Change*, Cambridge, MA: MIT Press.

——, Hughes, T. P., and Pinch, T.J. (eds), *The Social Construction of Technological Systems: New Directions in the Sociology and History of Technology*, Cambridge, MA: MIT Press, 1987.

Bloor, D., *Knowledge and Social Imagery*, London: Routledge & Kegan Paul, 1976.

Bowler, P. J., *The Non-Darwinian Revolution: Reinterpreting a Historical Myth*, Baltimore and London: The Johns Hopkins University Press, 1988.

Bradley, I., *The Annotated Gilbert and Sullivan*. Harmondsworth: Penguin Books, 1982.

Brignall, V., 'The eugenics movement Britain wants to forget', *New Statesman*, 9 December 2010, available at http://www.newstatesman.com/society/2010/12/british-eugenics-disabled, accessed 23 February 2014.

Bud, R., 'Framed in the public sphere: Tools for the conceptual history of "applied science" – a review paper', *History of Science*, 2013, 51: 413–433.

—— 'Making sense of modernity: the categories of pure and applied science in the public sphere of early twentieth-century Britain', Paper presented at the British Society for the History of Science annual conference, St Andrews, UK, 2014.

—— and Roberts, G.K., *Science Versus Practice: Chemistry in Victorian Britain*, Manchester and Dover, NH: Manchester University Press, 1984.

Bulmer, M., *Francis Galton: Pioneer of Heredity and Biometry*, Baltimore and London: The Johns Hopkins University Press, 2003.

Burney, I., *Poison, Detection, and the Victorian Imagination*, Manchester: Manchester University Press, 2006.

—— 'Our environment in miniature: Dust and the early twentieth-century forensic imagination', *Representations*, 2013, 121 (1): 31–59.

—— and Pemberton, N., 'The rise and fall of celebrity pathology', *British Medical Journal*, 2010, 341: 1319–1321.

—— and Pemberton, N., 'Bruised witness: Bernard Spilsbury and the performance of early twentieth-century English forensic pathology', *Medical History*, 2011, 55: 41–60.

—— and Pemberton, N. 'Making space for criminalistics: Hans Gross and fin-de- siècle CSI', *Studies in History and Philosophy of Biological and Biomedical Sciences*, 2013, 44 (1): 16–25.

Butterfield, H., *The Whig Interpretation of History*, London: G. Bell and Sons 1950 (original edition 1931).

Campbell, T., *Wireless Writing in the Age of Marconi*, Minneapolis: University of Minnesota Press, 2006.

Caplan, J. and Torpey, J. (eds), *Documenting Individual Identity: The Development of State Practices in the Modern World*, Princeton and Oxford: Princeton University Press, 2001.

—— and Torpey, J., 'Introduction', in J. Caplan and J. Torpey (eds), *Documenting Individual Identity: The Development of State Practices in the Modern World*, Princeton and Oxford: Princeton University Press, 2001, 1–13.

Cardwell, D.S.L., *The Organisation of Science in England: A Retrospect*, Melbourne, London, Toronto: Wm Heinemann Ltd, 1957.

Caudill, D.S. 'Arsenic and old chemistry: Images of mad alchemists, experts attacking experts, and the crisis in forensic science', *Villanova University School of Law Working Paper Series, Paper 136, 2009*, available at http://digitalcommons.law.villanova.edu/cgi/viewcontent.cgi?article=1140&context=wps, accessed 3 January 2015.

—— *Stories About Science in Law: Literary and Historical Images of Acquired Expertise*, Burlington, VT: Ashgate, 2011.

Chamberlain, D. C., 'Paper', in M. F. Suarez and H. R Woudhuysen (eds), *The Book: A Global History*, Oxford and New York: Oxford University Press, 16–129.

Charman, S. D. 'The forensic confirmation bias: A problem of evidence integration, not just evidence evaluation', *Journal of Applied Research in Memory and Cognition*, 2013, 2 (1): 56–58.

Chirnside, R. C. and Hammence, J. H., *The 'Practising Chemists' A History of the Society for Analytical Chemistry 1874–1974*, London: The Society for Analytical Chemistry, 1974.

Clarke, S. 'Pure science with a practical aim: The meanings of fundamental research in Britain, circa 1916–1950', *Isis*, 2010, 101 (2): 285–311.

Cole, S.A. *Suspect Identities: A History of Fingerprinting and Criminal Identification*, Cambridge, MA: Harvard University Press, 2001.

—— and Dioso-Villa, R., 'Investigating the "*CSI* effect": Media and litigation crisis in criminal law', *Stanford Law Review*, 2009, 61 (6): 1335–1373.

Collins, H. M., 'The TEA set: Tacit knowledge and scientific networks', *Science Studies*, 1974, 4: 165–186.

—— *Changing Order: Replication and Induction in Scientific Practice*, London: Sage, 1983.

—— *Tacit and Explicit Knowledge*, Chicago and London: University of Chicago Press, 2010.

Corfield, P., *Power and the Professions in Britain 1700–1850*, London and New York: Routledge, 1995.

Critchley, T. A., *A History of Police in England and Wales 900–1966*, London: Constable, 1967.

Crowther, M. A. and White, B. *On Soul and Conscience. The Medical Expert and Crime. 150 Years of Forensic Medicine in Glasgow*, Aberdeen: Aberdeen University Press, 1988.

Cullen, T., *Crippen: The Mild Murderer*, London: Bodley Head, 1977.

Daston, L. and Galison, P., 'The image of objectivity', *Representations*, 1992, 40: 81–128.

—— *Objectivity*, New York: Zone Books, 2007.

Davie, N. *Tracing the Criminal: The Rise of Scientific Criminology in Britain, 1860–1918*, Oxford: Bardwell Press, 2005.

Desrosières, A., *The Politics of Large Numbers: A History of Statistical Reasoning*. Cambridge, MA: Harvard University Press, 1998.

Dixon, T., *Science and Religion: A Very Short Introduction*, Oxford: Oxford University Press, 2008.

Donaldson, N., *In Search of Dr Thorndyke: The Story of R. Austin Freeman's Great Scientific Investigator and His Creator*, Bowling Green, OH: Bowling Green University Popular Press, 1971.

Du Boff, R. B., 'The rise of communications regulation: The telegraph industry, 1844–1880', *Journal of Communication*, 1984, 34 (3): 52–66.

Duff, P., 'The Scottish criminal jury: A very peculiar institution', *Law and Contemporary Problems*, 1999, 62 (2): 173–201.

Duvall, N., *Forensic Medicine in Scotland, 1914–39*, unpublished PhD thesis, University of Manchester, 2013.

Dwyer, D. M., 'Expert evidence in the English civil courts, 1550–1800', *The Journal of Legal History*, 2007, 28 (1): 93–118.

Early, J. E., 'Technology, modernity, and 'the little man': Crippen's capture by wireless', *Victorian Studies*, 1996, 39 (3), 309–337.

Edgerton, D. 'The linear model did not exist', in K. Grandin, N. Worms, and S. Widmalm (eds), *The Science-Industry Nexus: History, Policy, Implications*, Sagamore Beach, MA: Science History Publications, 2004, 31–57.

—— *The Shock of the Old: Technology and Global History since 1900*, Place: London: Profile, 2006.

Edmond, G. (ed.), *Expertise in Regulation and Law*, Aldershot and Burlington, VT: Ashgate, 2004.

Evans, C., *Crime Scene Investigation (Criminal Investigations)*, New York: Chelsea House, 2009.

Feyerabend, P., *Against Method*, London: New Left Books, 1975.

Fisher, J., *Forensics Under Fire: Are Bad Science and Dueling Experts Corrupting Criminal Justice?* New Brunswick, NJ: Rutgers University Press, 2008.

Foran, D. R., Wills, B. E., Kiley, B. M., Jackson, C. B. and Trestrail III, J. H., 'The conviction of Dr. Crippen: New forensic findings in a century-old murder', *Journal of Forensic Sciences*, 2011, 56 (1): 233–240.

Forbes, T. R., *Surgeons at the Bailey: English Forensic Medicine to 1878*, New Haven, CT and London: Yale University Press, 1985.

Foucault, M., *Discipline and Punish: The Birth of the Prison* (trans. A. Sheridan), London: Allen Lane, 1977.

French, M. and Phillips, J., *Cheated Not Poisoned? Food Regulation in the United Kingdom, 1875–1938*, Manchester and New York: Manchester University Press, 2000.

Gardiner, D. and Sorley Walker, K. (eds), *Raymond Chandler Speaking*, Berkeley and Los Angeles: University of California Press, 1997.

Garland, D., 'The criminal and his science: A critical account of the formation of criminology at the end of the nineteenth century', *British Journal of Criminology*, 1985, 25 (2): 109–137.

—— 'British criminology before 1935', *British Journal of Criminology*, 1988, 28 (2): 1–17.

Giannelli, P. C., 'The 2009 NAS Forensic Science Report: A literature review', *Criminal Law Bulletin*, 2012, 48 (2): 378–393.

Gilberg, M., 'Alfred Lucas: Egypt's Sherlock Holmes', *Journal of the American Institute for Conservation*, 1997, 36 (1): 31–48.

Godfrey, B., *Crime in England 1880–1945: The Rough and the Criminal, the Policed and the Incarcerated*, Abingdon and New York: Routledge, 2014.

Godin, B., 'The linear model of innovation: The historical construction of an analytical framework', *Science Technology and Human Values*, 2006, 31 (6): 639–667.

Golan, T., 'The history of scientific expert testimony in the English courtroom', *Science in Context*, 1999, 12 (1): 7–32.

—— *Laws of Men and Laws of Nature: The History of Scientific Expert Testimony in England and America*, Cambridge, MA and London: Harvard University Press, 2004.

—— 'Revisiting the history of scientific expert testimony', *Brooklyn Law Review*, 2008, 73 (3): 879–942.

Gooday, G., 'Liars, experts and authorities', *History of Science*, 2008, 46 (4): 431–456.

—— 'Placing or replacing the laboratory in the history of science?', *Isis*, 2008, 99 (4): 783–795.

—— ' "Vague and artificial": The historically elusive distinction between pure and applied science', *Isis*, 2012, 103 (3): 546–554.

Goodman, J., *The Crippen File*, London and New York: Allison & Busby: 1985.

Green, M. A., 'Is Sir Bernard Spilsbury dead?' in A. R. Brownlie (ed.), *Criminal Investigation Art or Science?* Edinburgh and London: Scottish Academic Press, 1984, 23–26.

Grimes, H., *The Late Victorian Gothic: Mental Science, the Uncanny, and Scenes of Writing*, Farnham and Burlington, VT: Ashgate, 2011.

Hamlin, C. (1986) 'Scientific method and expert witnessing: Victorian perspectives on a modern problem', *Social Studies of Science*, 16 (3): 485–513.

Hammond, P. W. and Egan, H., *Weighed in the Balance: A History of the Laboratory of the Government Chemist*, London: HMSO, 1992.

Hansen, K. V. 'Historical sociology and the prism of biography: Lillian Wineman and the trade in Dakota beadwork, 1893–1929', *Qualitative Sociology*, 1999, 22 (4): 353–368.

Harrison, M., *In the Footsteps of Sherlock Holmes*, Newton Abbot: David and Charles, 1971.

Harvey, E. and Derksen, L., 'Science fiction or social fact? An exploratory content analysis of popular press reports of the *CSI* effect', in M. Byers and V. M. Johnson (eds), *The CSI Effect: Television, Crime and Governance*, 2009, Lanham, MD: Lexington Books, 3–28.

Haynes, R., *From Faust to Strangelove: Representations of the Scientist in Western Literature*, Baltimore, MD: The Johns Hopkins University Press, 1994.

—— 'From alchemy to artificial intelligence: Stereotypes of the scientist in Western literature', *Public Understanding of Science*, 2003, 12 (3): 243–253.

Higgs, E. *Identifying the English: A History of Personal Identification 1500 to the Present*, London and New York: Continuum, 2011.

Hilton, M., *Smoking in British Popular Culture 1800–2000: Perfect Pleasures*, Manchester and New York: Manchester University Press, 2000.

Hindmarsh, R. and Prainsack, B. (eds), *Genetic Suspects: Global Governance of DNA Profiling and Databasing*, Cambridge: Cambridge University Press, 2010.

Hobsbawm, E. J., *Bandits*, Harmondsworth: Pelican 1972.

Horn, D. G., 'Making criminologists: Tools, techniques, and the production of scientific authority', in P. Becker and. R. F. Wetzell (eds), *Criminals and Their Scientists: The History of Criminology in International Perspective*, Cambridge and New York: Cambridge University Press, 2006, 317–336.

Horrocks, S., *Consuming Science: Science, Technology and Food in Britain, 1870–1939*, unpublished PhD thesis, University of Manchester, 1993.

Huber, P., *Galileo's Revenge: Junk Science in the Courtroom*, New York: Basic, 1991.

Hunter, M., *The Royal Society and Its Fellows, 1660–1700: The Morphology of an Early Scientific Institution*, Chalfont St Giles, Bucks: British Society for the History of Science, 1982.

—— *Robert Boyle 1627–91: Scrupulosity and Science*, Woodbridge and Rochester, NY: Boydell Press, 2000.

Hutchings, P. J. 'Modern forensics: Photography and other suspects', *Cardozo Studies in Law and Literature*, 1997, 9 (2): 229–243.

Huws, U., *Your Job in the Eighties: Woman's Guide to New Technology*, London: Pluto, 1982.

Jasanoff, S., *Science at the Bar: Law, Science and Technology in America*. Cambridge, MA and London: Harvard University Press, 1995.

Jenkins, P., 'Varieties of Enlightenment criminology: Beccaria, Godwin, de Sade', *British Journal of Criminology*, 1984, 24 (2): 112–130.

Jensen Wallach, J., 'Building a bridge of words: The literary autobiography as historical source material', *Biography*, 2006, 29 (3): 446–461.

Jones, C.A.G., *Expert Witnesses: Science, Medicine, and the Practice of Law*, Oxford, Clarendon Press, 1994.

Joseph, A. M., 'Anthropometry, the police expert, and the Deptford murders: The contested introduction of fingerprinting for the identification of criminals in late Victorian and Edwardian Britain', in J. Caplan and J. Torpey (eds), *Documenting Individual Identity: The Development of State Practices in the Modern World*, Princeton and Oxford, Princeton University Press, 2001, 164–183.

Joyce, C., 'The clues to forensic science found by computer', *New Scientist*, 27 September 1984, 8.

Kaluszynski, M. 'Republican identity: Bertillonage as government technique', in J. Caplan and J. Torpey (eds), *Documenting Individual Identity: The Development of State Practices in the Modern World*, Princeton and Oxford, Princeton University Press, 2001, 123–138.

Karstedt, S., 'Strangers, mobilisation and the production of weak ties: Railway traffic and violence in nineteenth-century South-West Germany', in B. S. Godfrey, C. Emsley and G. Dunstall (eds), *Comparative Histories of Crime*, Devon: Willan, 2003, 89–109.

Kestner, J. A., *Sherlock's Men: Masculinity, Conan Doyle and Cultural History*, Aldershot and Brookfield USA: Ashgate, 1997.

—— *The Edwardian Detective, 1901–1915*, Aldershot and Brookfield USA: Ashgate, 2000.

Kincheloe, J. L. and Berry, K. S., *Rigour and Complexity in Educational Research: Conceptualizing the Bricolage*, Maidenhead: Open University Press, 2004.

Knepper, P., *The Invention of International Crime: A Global Issue in the Making, 1881–1914*, Basingstoke and New York: Palgrave Macmillan, 2010.

Krinsky, C., *The Ashgate Research Companion to Moral Panics*, Farnham and Burlington, VT: Ashgate, 2013.

Kuhn, T. S., *The Structure of Scientific Revolutions*, Chicago, IL: University of Chicago Press, 1962.

Landsman, S., 'One hundred years of rectitude: Medical witnesses at the Old Bailey, 1717–1817', *Law and History Review*, 1998, 16 (3): 445–494.

Larson, E., *Thunderstruck*, New York: Broadway Books, 2007.

Latour, B., *Science in Action*, Cambridge: Harvard University Press, 1987.

Laybourn, K. and Taylor, D., *Policing in England and Wales, 1918–39: The Fed, Flying Squads and Forensics*, Basingstoke and New York: Palgrave Macmillan, 2011.

Law, J., *After Method: Mess in Social Science Research*, London: Routledge, 2004.

Lawrence, C., 'Incommunicable knowledge: Science, technology and the clinical art in Britain 1850–1914', *Journal of Contemporary History*, 1985, 20 (4): 503–520.

Lindsey, C., *Sherlock Holmes and a Question of Science*, Leatherhead: Hadley Pager Info, 2006.

Locke, J., *From Ghouls to Gangsters: The Career of Arthur B. Reeve*, Elkhorn, CA: Off-Trail Publications, 2007, 2 vols.

Lowrey, B., 'The timelessness of Stephen Potter's gamesmanship', *The Virginia Quarterly Review: A National Journal of Literature and Discussion*, 1993, 69 (4): 718–726.

Lynch, M. J., 'The Power of oppression: Understanding the history of criminology as a science of oppression', *Critical Criminology*, 2000, 9 (1/2): 144–152.

Lynch, M. 'Science, truth, and forensic cultures: The exceptional legal status of DNA evidence', *Studies in History and Philosophy of Biological and Biomedical Sciences*, 2013, 44 (1): 60–70.

—— Cole, S. A., McNally, R. and Jordan, K., *Truth Machine: The Contentious History of DNA Fingerprinting*, Chicago, IL and London: University of Chicago Press, 2008.

McFarland, M., *Chronology of Key Events in the History of the Daily Mail*, 2013, available at http://gdc.gale.com/assets/files/daily_mail/chronology_of_notable_events.pdf, accessed 27 August 2014.

Machado, H. and Prainsack, B., *Tracing Technologies: Prisoners' Views in the Era of CSI*, Fanham and Burlington, VT: Ashgate, 2012.

MacKenzie, D. and Wajcman, J. (eds), *The Social Shaping of Technology: How the Refrigerator Got its Hum*, Milton Keynes: Open University Press, 1985.

MacKenzie, F. A, *The Mystery of the Daily Mail: 1896–1921*. London: Associated Newspapers, 1921.

Macrakis, K., *Prisoners, Lovers, and Spies: The Story of Invisible Ink from Herodotus to al-Qaeda*, New Haven, CT and London: Yale University Press, 2014.

McFarland, M., *Chronology of Key Events in the History of the Daily Mail*, 2013, available at http://gdc.gale.com/assets/files/daily_mail/chronology_of_notable_events.pdf, accessed 27 August 2014.

McWilliam, R., *The Tichborne Claimant. A Victorian Sensation*, London: Hambledon Continuum, 2007.

Marvin, C., *When Old Technologies Were New: Thinking About Electric Communication in the Late Nineteenth Century*, New York and Oxford: Oxford University Press, 1988.

Mazévet, M. *Edmond Locard: Le Sherlock Holmes Français*, Lyon: Editions des Traboule, 2006.

Mnookin, J. L., 'Scripting expertise: The history of handwriting identification evidence and the judicial construction of reliability', *Virginia Law Review*, 2001, 87 (8): 1723–1845.

—— 'Idealizing science and demonizing experts: An intellectual history of expert evidence', *Villanova Law Review*, 2007, 52 (101): 763–801.

Morrison, W., *Criminology, Civilisation and the New World Order*, Abingdon and New York: Routledge Cavendish, 2006.

Muhlbacher, T, 'Hans Gross, the father of criminology, and Arthur Conan Doyle', in G. Heuer (ed), *Sexual Revolutions: Psychoanalysis, History and the Father*, London and New York: Routledge, 2010, 55–64.

Muka, S. K., 'Portrait of an outsider: Class, gender, and the scientific career of Ida M. Mellen', *Journal of the History of Biology*, 2013, available at http://link.springer.com/article/10.1007/s10739–013–9354-z, accessed 20 April 2013.

Nagel T., *A View from Nowhere*, Oxford: Oxford University Press, 1986.

National Research Council, National Academy of Sciences (NRCNAS), *Strengthening Forensic Science in the United States: A Path Forward*, Washington DC, National Academies Press, 2009.

Narayanan A. and Bennun, M. (eds), *Law, Computer Science and Artificial Intelligence*, New York: Ablex, 1991.

Newhall, B., 'History of photography: Photojournalism', *Encyclopedia Britannica*, 2014, available at http://www.britannica.com/EBchecked/topic/457919/history-of-photography/252857/Photojournalism, accessed 27 August 2014.

Nocks, L., 'T. H. Huxley: The evolution of the bulldog', in B. Regal (ed), *Icons of Evolution: An Encyclopedia of People, Evidence, and Controversies*, Volume 1, Westport, CT: ABC-CLIO/Greenwood, 2008, 57–86.

Novak, D., *Realism, Photography and Nineteenth-Century Fiction*, Cambridge: Cambridge University Press, 2008.

Nye, M. J., 'Scientific biography: History of science by another means?, *Isis*, 2006, 97 (2): 322–329.

O'Brien, J. F., *The Scientific Sherlock Holmes: Cracking the Case with Science and Forensics*, Oxford and New York: Oxford University Press, 2013.

Oddy, D. J., 'Food quality in London and the rise of the public analyst, 1870–1939', in P. J. Atkins, P. Lummel and D. J. Oddy (eds), *Food and the City in Europe since 1800*, Farnham and Burlington, VT: Ashgate, 2007, 91–103.

Oldfield, R., *Outrage: The Edalji Five and the Shadow of Sherlock Holmes*, Cambridge: Vanguard Press, 2010.

Osterburg, J. W. and Ward, R. W., *Criminal Investigation: A Method for Reconstructing the Past*, Waltham, MA: Anderson, 2014, 7th edition.

Panek, L., *The Origins of the American Detective Story*, Jefferson, NC: MacFarland & Co., 2006.

Pavlich, G., 'The subjects of criminal identification', *Punishment & Society*, 2009, 11 (2): 171–190.

Perkin, H., *The Rise of Professional Society: England Since 1880*, London and New York: Routledge, 1989.

Peterson, M. J., *The Medical Profession in Mid-Victorian London*, Berkeley and Los Angeles: University of California Press, 1978.

Piazza, P., *Aux Origines de la Police Scientifique: Alphonse Bertillon, Precurseur de la science du crime*, Clamecy, France: Karthala, 2011.

Popkin, J. D., *History, Historians & Autobiography*, Chicago, IL and London: University of Chicago Press, 2005.

Prainsack, B. and Kitzberger, M., 'DNA behind bars: Other ways of knowing forensic DNA technologies', *Social Studies of Science*, 2009, 39 (1): 51–79.

Pugh, M., *We Danced All Night: A Social History of Britain Between the Wars*, London: Vintage, 2009.

Radhakrishna, M., 'Colonial construction of a "criminal" tribe: Yerukulas of Madras Presidency', *Economic and Political Weekly*, 2000, 35 (28/29): 2553–2563.

—— *Dishonoured by History: 'Criminal Tribes' and British Colonial Policy*, Hyderabad: Orient Longman, 2001.

Rafter, N. H., *Creating Born Criminals*, Urbana and Chicago: University of Illinois Press, 1997.

Rawlings, P., *Policing: A Short History*, Abingdon and New York: Routledge, 2014 (first published 2002 by Willan Publishing).

Richards, A. J. and Gore, B., *Holmes, Chemistry and the Royal Institution*, London: Sherlock Holmes Society of London, 1998.

—— 'Sherlock Holmes: A study in chemistry' in A. J. Richards and B. Gore, *Holmes, Chemistry and the Royal Institution*, London: Sherlock Holmes Society of London, 1998, 23–34.

Rose, A., *Lethal Witness: Sir Bernard Spilsbury Honorary Pathologist*, Chalfont, Stroud: Sutton, 2007.

—— 'Lethal witness', paper presented at the Royal Society of medicine, 2 April 2009, available at netk.net.au/Forensic/Spilsbury1.pdf, accessed 2 January 2014.

Rossiter, M. W., *Women Scientists in America: Struggles and Strategies to 1940, Vol. 1*. Baltimore, MD: The Johns Hopkins University Press, 1982.

Rowbotham, J. and Stevenson, K. (eds), *Criminal Conversations: Victorian Crimes, Social Panic and Moral Outrage*, Columbus: Ohio State University Press, 2005.

——, Stevenson, K. and Pegg, S., *Crime News in Modern Britain: Press Reporting and Responsibility: 1820–2010*, Houndmills, Basingstoke: Palgrave Macmillan, 2013.

Russell, C. A, Cole, N. G. and Roberts, G. K., *Chemists by Profession: The Origins and Rise of the Royal Institute of Chemistry*, Milton Keynes: Open University Press/Royal Institute of Chemistry, 1977.

Russell, N., 'Science and scientists in Victorian and Edwardian literary novels: Insights into the emergence of a new profession', *Public Understanding of Science*, 2007, 16 (2): 205–222.

—— 'The new men: Scientists at work in popular British fiction between the early 1930s and the late 1960s', *Science Communication*, 2009, 31 (1): 29–56.

Ryan, J. R., *Picturing Empire: Photography and the Visualization of the British Empire*, Chicago IL: Chicago University Press, 1997.

Rzepka, C. J., *Detective Fiction*, Cambridge and Malden, MA: Polity, 2005.

Saks, M. J., 'Banishing ipse dixit: The impact of Kumho Tire on forensic identification science', *Washington and Lee Law Review*, 2000, 57 (3): 879–900, available at http://scholarlycommons.law.wlu.edu/wlulr/vol57/iss3/7, accessed 6 January 2015.

Schatzberg, E., 'From art to applied science', *Isis*, 2012, 103 (3): 555–563.

Schwarz, H., *Constructing the Criminal Tribe in Colonial India: Acting Like a Thief*, Chichester: John Wiley, 2010.

Science and Technology Committee, *Forensic Science on Trial: Seventh Report of Session 2004–05*, London, House of Commons, 2005, available at http://www.publications.parliament.uk/pa/cm200405/cmselect/cmsctech/96/96i.pdf, accessed 12 January 2015.

Seal, G., 'The Robin Hood principle: Folklore, history and the social bandit', *Journal of Folklore Research*, 2009, 46 (1): 67–89.

Sekula, A., 'The body and the archive', *October*, 1986, 39 (Winter): 3–64.

Sengoopta, C., *Imprint of the Raj: How Fingerprinting Was Born in Colonial India*, Basingstoke and Oxford: Pan, 2003.

Shapin, S., *A Social History of Truth: Civility and Science in Seventeenth Century England*, Chicago, IL: Chicago University Press, 1994.

—— *The Scientific Revolution*, Chicago, IL and London: University of Chicago Press, 1996.

—— *The Scientific Life: A Moral History of a Late Modern Vocation*, Chicago, IL and London: University of Chicago Press, 2008.

—— and Schaffer, S., *Leviathan and the Air Pump: Hobbes, Boyle and the Experimental Life*, Princeton, NJ: Princeton University Press, 1985.

Shapiro, B., *Probability and Certainty in Seventeenth-Century England: A Study of the Relationships Between Natural Science, Religion, History, Law and Literature*, Princeton, NJ: Princeton University Press, 1983.

—— *Beyond Reasonable Doubt and Probable Cause: Historical Perspectives on the Anglo-American Law of Evidence*, Berkeley and Los Angeles, CA: University of California Press, 1991.

—— *A Culture of Fact: England 1550–1720*, Ithaca, NY and London: Cornell University Press, 2000.

—— '"Fact" and the proof of fact in Anglo-American law (c. 1500–1850)', in A. Sarat, L. Douglas and M. Umphrey (eds), *How Law Knows*, Stanford, CA: Stanford University Press, 2007, 25–71.

—— and Frank, Jr, R. G., *English Scientific Virtuosi in the 16th and 17th Centuries*, Los Angeles, CA: William Andrews Clark Memorial Library, 1979.

Shpayer-Makov, H., 'Explaining the rise and success of detective memoirs in Britain', in C. Emsley and H. Shpayer-Makov (eds), *Police Detectives in History, 1750–1950*, Aldershot and Burlington, VT: Ashgate, 2006, 103–133.

—— *The Ascent of the Detective: Police Sleuths in Victorian and Edwardian England*, Oxford and New York: Oxford University Press, 2011.

—— 'Shedding the uniform and acquiring a new masculine image: the case of the late-Victorian and Edwardian police detective', in D. G. Barrie and S. Broomhall (eds.), *A History of Police and Masculinities, 1700–2010*, London and New York: Routledge, 2012, 141–162.

Sismondo, S. *An Introduction to Science and Technology Studies*, Malden, MA: Blackwell Publishing, 2010.

Sleigh, C. *Literature and Science*, Houndmills, Basingstoke: Palgrave Macmillan, 2011.

Smith, R., 'Forensic pathology, scientific expertise, and the criminal law', in R. Smith and B. Wynne (eds), *Expert Evidence: Interpreting Science in the Law*, London and New York: Routledge, 1989, 56–92.

Smith, S. and Watson, J., *Reading Autobiography: A Guide for Interpreting Life Narratives*, Minneapolis: University of Minnesota Press, 2001.

Söderqvist, T., 'What is the use of writing lives of recent scientists?', in R. E. Doel and T. Söderqvist (eds), *The Historiography of Contemporary Science, Technology and Medicine: Writing Recent Science*, Abingdon and New York: Routledge, 2006, 99–127.

Standage, T., *The Victorian Internet: The Remarkable Story of the Telegraph and the Nineteenth Century's Online Pioneers*, London: Phoenix, 1999.

Stanziani, A. and Atkins, P. J. 'From laboratory expertise to litigation: The Municipal Laboratory of Paris and the Inland Revenue Laboratory in London, 1870–1914: A comparative analysis', in C. Rabier (ed), *Fields of Expertise: A Comparative History of Expert Procedures in Paris and London, 1600 to Present*, Newcastle upon Tyne: Cambridge Scholars Publishing, 2007, 317–338.

Steenberg, L., *Forensic Science in Contemporary American Popular Culture: Gender, Crime, and Science*, New York and Abingdon: Routledge, 2013.

Steere-Williams, J., 'A conflict of analysis: analytical chemistry and milk adulteration in Victorian Britain', *Ambix*, 2014, 61 (3): 279–298.

Steinert, H., 'Fin de siècle criminology', *Theoretical Criminology*, 1997, 1 (1): 111–129.

Strauss, C., 'The imaginary', *Anthropological Theory*, 2006, 6 (3): 322–344.

Sweeney, S. E., 'The magnifying glass: Spectacular distance in Poe's "Man of the Crowd" and beyond', *Poe Studies/Dark Romanticism*, 2003, 36 (1–2): 3–17.

Taylor, G., *Forensic Enforcement: The Role of the Public Analyst*, Cambridge: RSC Publishing, 2010.

Thomas, R. R., *Detective Fiction and the Rise of Forensic Science*, Cambridge and New York: Cambridge University Press, 1999.

Turner, F. M., *Between Science and Religion: The Reaction to Scientific Naturalism in Late Victorian England*, New Haven, CT: Yale University Press, 1974.

Van Dover, J. K., *You Know My Method: The Science of the Detective*, Bowling Green, OH: Bowling Green State University Popular Press, 1994.

Van Wyhe, J., *Phrenology and the Origins of Victorian Scientific Naturalism*, Aldershot and Burlington, VT: Ashgate, 2004.

Vyleta, D. M, *Crime, Jews and News: Vienna 1895–1914*, New York and Oxford: Berghahn, 2007.

Wagner, E. J., *The Science of Sherlock Holmes From Baskerville Hall to the Valley of Fear, the Real Forensics Behind the Great Detective's Greatest Cases*, Hoboken, NJ: John Wiley & Sons, 2006.

Wagner, K. A., *Thuggee: Banditry and the British in Early Nineteenth-Century India*, Houndmills, Basingstoke, Palgrave Macmillan, 2007.

Wajcman, J., *Feminism Confronts Technology*, Cambridge: Polity, 1991.

Wallach, J. J., 'Building a bridge of words: The literary autobiography as historical source material', *Biography*, 2006, 29 (3): 446–461.

Ward, J., *Origins and Development of Forensic Medicine and Forensic Science in England, 1823–1946*, unpublished PhD thesis, The Open University, 1993.

Ward, T., 'A mania for suspicion: Poisoning, science and the law', in J. Rowbotham and K. Stevenson (eds), *Criminal Conversations: Victorian Crime, Social Panic, an Moral Outrage*, Columbus: Ohio State University Press, 2005, 140–156.

Warlow, T. A., *Firearms, the Law and Forensic Ballistics*, Boca Raton, FL: CRC Press, 2005.

Watson, K. D., 'The chemist as expert: The consulting career of Sir William Ramsay', *Ambix*, 1995, 42 (3): 143–159.

—— *Poisoned Lives: English Poisoners and Their Victims*, London and New York: Hambledon and London, 2004.

—— 'Medical and chemical expertise in English trials for criminal poisoning, 1750–1914', *Medical History*, 2006, 50: 373–390.

—— *Dr Crippen*, Kew, Richmond: The National Archives, 2007.

—— *Forensic Medicine in Western Society: A History*, Abingdon: Routledge, 2011.

Wayman, J., Jain, A., Maltoni, D., and Maio, D. (eds), *Biometric Systems: Technology, Design and Performance Evaluation*, London: Springer-Verlag, 2005.

Wiener, M. J. 'Murderers and "reasonable men": The "criminology" of the Victorian judiciary', in P. Becker and R. F. Wetzell (eds), *Criminals and Their Scientists: The History of Criminology in International Perspective*, Cambridge and New York: Cambridge University Press, 2006, 43–60.

Williams, R. and Johnson, P., *Genetic Policing: The Use of DNA in Criminal Investigations*, Cullompton: Willan Publishing, 2008.

Willis, M., *Literature and Science: A Reader's Guide to Essential Criticism*, London and New York: Palgrave, 2015.

Wonder, A.K.Y., 'Science and law, a marriage of opposites', *Journal of the Forensic Science Society*, 1989, 29 (2): 75–76.

Youngson, A. J., *The Scientific Revolution in Victorian Medicine*, London: Croom Helm, 1979.

Index